Hydrogen and other Alternative Fuels for Air and Ground Transportation

Hydrogen and other Alternative Fuels for Air and Ground Transportation

Edited by

H.W. Pohl
Daimler-Benz Aerospace Airbus, Hamburg, Germany

JOHN WILEY & SONS
Chichester · New York · Brisbane · Toronto · Singapore

Publication EUR 15898 EN of the
European Commission,
Directorate-General XII – Science, Research and Development, Brussels

© ECSC-EEC-EAEC, Brussel-Luxembourg, 1995.

LEGAL NOTICE
Neither the European Commission nor any person acting on behalf of the Commission is responsible for the use which might be made of the following information.

Published in 1995 by John Wiley & Sons Ltd,
                Baffins Lane, Chichester,
                West Sussex PO19 1UD, England

Telephone: *National*   01243 779777
             International (+44) 1243 779777

All rights reserved.

No part of this book may be reproduced by any means, or transmitted, or translated into a machine language without the written permission of the publisher.

The figures in Chapter 4 have been reprinted with permission from, *Renewable Energy: Sources for Fuels and Electricity*, by Johansson Kelly, Roddy Williams and Burnham, published by Island Press, Washington D.C. and Covelo, California © 1993 Island Press.

Other Wiley Editorial Offices

John Wiley & Sons, Inc., 605 Third Avenue,
New York, NY 10158-0012, USA

Jacaranda Wiley Ltd, 33 Park Road, Milton,
Queensland 4064, Australia

John Wiley & Sons (Canada) Ltd, 22 Worcester Road,
Rexdale, Ontario M9W 1L1, Canada

John Wiley & Sons (SEA) Pte Ltd, 37 Jalan Pemimpin #05-04,
Block B, Union Industrial Building, Singapore 2057

Library of Congress Cataloging-in-Publication Data

Hydrogen and other alternative fuels for air and ground transportation
  edited by H. W. Pohl.
    p. cm
  Proceedings of a workshop held in Brussels in June 1993.
  Includes bibliographical references and index.
  ISBN 0 471 95336 9
  1. Hydrogen as fuel — Congresses.   I. Pohl, H. W.
TP359.H8H843   1995
665.8'1 — dc20   95-17785
                                                                          CIP

British Library Cataloguing-in-Publication Data

A catalogue record for this book is available from the British Library

ISBN 0 471 95336 9

Typeset from author's disk in 10/12pt Times by Keytec Typesetting, Bridport
Printed and bound in Great Britain by Biddles Ltd, Guildford and King's Lynn
This book is printed on acid-free paper responsibly manufactured from sustainable forestation, for which at least two trees are planted for each one used for paper production.

Contents

Preface	ix
Summary	xi
List of Authors	xiii

1. Long-term Trends in Energy Consumption and in Availability and Cost of Conventional Fuel — 1

- 1.1 Introduction — 1
- 1.2 Trends in global energy requirements — 1
- 1.3 Trends in fuel requirements for air and road transportation — 4
- 1.4 Energy reserves and resources — 7
- 1.5 Price trends for conventional fuels — 10
- 1.6 Availability of alternative fuels — 12
- 1.7 Adoption of alternative fuels for air and road transport — 14
- 1.8 Summary — 16
- 1.9 References — 18

2. Energy and the Environment — 19

- 2.1 Introduction — 19
- 2.2 Problems from fossil fuels — 23
- 2.3 Alternative energy production and fuels — 29
- 2.4 Conclusions — 32
- 2.5 References — 33

3. Characteristics of Alternative Fuels – Limiting the Choice — 35

- 3.1 Introduction — 35
- 3.2 Energy sources and energy carriers — 35
- 3.3 Characteristics of energy carriers — 36
- 3.4 Criteria for selection of energy carrier — 37
- 3.5 Selection of energy carrier for aircraft — 38
- 3.6 Selection of energy carrier for road vehicles — 39

4. Clean Energy Sources — 41
   4.1 Introduction — 41
   4.2 Review of renewable energy potentials — 43
   4.3 Major development trends and R&TD underway — 55
   4.4 Trends in efficiency/economics — 61
   4.5 Conclusions and recommendations — 67
   4.6 References — 71

5. Technology for Gaseous Hydrogen Production — 75
   5.1 Current state of H_2 production — 75
   5.2 Liquid hydrogen — 76
   5.3 Integration of hydrogen supply for vehicles and planes into the electric power grid — 76
   5.4 Decomposition of natural gas into hydrogen and pure carbon — 79
   5.5 Concept of hydrogen production in decentralized installations — 81

6. Technology for Cryofuel Production — 85
   6.1 Production processes used today — 85
   6.2 Processes under R&TD — 89
   6.3 Price trends — 90
   6.4 Storage and distribution — 96
   6.5 Safety record — 99
   6.6 Regulations — 103
   6.7 Conclusions — 104
   6.8 Recommendations — 104
   6.9 Experience from LH_2 space application — 104

7. Experience from LH_2 Space Application — 111
   7.1 Introduction — 111
   7.2 Rocket engine testing — 111
   7.3 Test facilities — 115
   7.4 Experience — 117
   7.5 Components applicable in air/ground transportation — 119
   7.6 Recommendations — 121

8. Experience from the TU-155 Experimental Aircraft — 123
   8.1 Introduction — 123
   8.2 The Tupolev Company's philosophy using hydrogen in transportation — 124
   8.3 The TU-155 experimental aircraft — 126
   8.4 Problems to be solved in cryogenic fuel system operation, engine supply and refuelling — 130
   8.5 Conclusions and proposals from the TU-155 experiences — 133
   8.6 Outlook for the future of cryogenic aviation in Russia — 134

9. Aeroengines for Alternative Fuels — 137
- Abstract — 137
- 9.1 Introduction — 137
- 9.2 Effect on airframe, engine and infrastructure — 140
- 9.3 Cryogenic engine design and operation — 143
- 9.4 Present activities and outlook — 151
- 9.5 Summary — 154
- 9.6 References — 155

10. Alternative Fuels in Aviation — 157
- 10.1 Summary of previous studies and experimental work on alternative fuels in aviation — 157
- 10.2 Current studies at Deutsche Aerospace Airbus — 161
- 10.3 Critical components and R&TD requirements — 174

11. Alternative Fuels in Ground Transportation — 177
- 11.1 Introduction — 177
- 11.2 Description of the hydrogen propulsion concept with internal combustion engines — 178
- 11.3 State of hydrogen vehicles — 183
- 11.4 Costs — 185
- 11.5 Reasons for the introduction of hydrogen technology and obstacles still overcome — 186
- 11.6 Evaluation of the application potential — 186

12. Phase II and Phase III.0 of the 100 MW Euro-Québec Hydro-Hydrogen Pilot Project (EQHHPP) — 189
- 12.1 Introduction — 189
- 12.2 The concept — 190
- 12.3 Milestones — 190
- 12.4 Phase II of the EQHHPP — 191
- 12.5 Economics — 195
- 12.6 Environmental aspects — 196
- 12.7 Phase III.0 — 197
- 12.8 Funding — 199
- 12.9 Outlook — 199
- 12.10 Acknowledgement — 200
- 12.11 References — 200

Preface

The transport sector is one of the major consumers of energy, accounting for 30% of global energy demand. Moreover, transportation is the fastest growing sector of worldwide economy, thus even increasing its share in future total energy use. The discussion on environmental phenomena, like global warming, and on the availability of fossil energy sources therefore have led to reflection on the use of alternative fuels in air and ground transportation. The long lead time for the implementation of any alternative fuel in the transport systems requires progress with enabling research activities.

In the light of the preparation for the 4th R&TD Framework Programme of the European Commission, a workshop was held in Brussels in June 1993 to discuss the rationale and the possibilities for alternative fuels for air and ground transportation and to identify possible actions which should be undertaken.

As a preparation for this workshop, reports have been written by leading experts from different industrial and research sectors covering future availability of conventional transportation fuels, environmental impacts of emissions from transportation, possible alternative fuels and the state of knowledge in alternative fuels production, handling and application.

Based on the state of knowledge represented by the papers compiled in this publication, and the discussions and findings of the workshop, a scenario regarding future air and ground transportation was developed and a possible European strategy to introduce alternative fuels into the transportation sector was formulated. These considerations along with proposals for European R&TD activities are subject of a study entitled

'Hydrogen and Other Alternative Fuels for Air and Ground Transportation'

which is available at Directorate General XII – Science, Research and Development – of the European Commission.

Summary

The majority of today's transportation fuels is based on crude oil. Predictions of population growth and scenarios on future energy demand indicate that the known reserves of crude oil will draw to an end between 2020 and 2040. Taking into account the total of fossil resources, i.e. conventional and unconventional oil (tar sands, oil shale), natural gas and coal, substitution of conventional hydrocarbon fuels in transportation is unlikely to be required during the 21st century for reasons of supply exhaustion alone.

However, producing gasoline, gasoil and kerosene from sources other than crude oil implies increasing emissions into the atmosphere and increasing production costs. Therefore, new energy carriers like hydrogen, whose production could be based on energy sources like wind-, hydro-, solar-power or biomass, may become economically competitive by say 2020 to 2030, even in absence of specific government action (selective taxation etc.) to force ahead the introduction of new fuels for the sake of protecting the atmosphere.

Regarding environmental aspects, carbon dioxide (CO_2) – responsible for 50% of the man-made greenhouse effect – and nitrogen oxides (NO_x) – adversely affecting the atmospheric ozone concentration – are the most crucial emissions from conventional fuels use. To avoid potentially grave climatic consequences, the increase in CO_2 content of the atmosphere must be stopped in the next decade. To stabilize the CO_2 concentration, a 60% reduction of emissions is required by the middle of the next century. Even if this drastic reduction is achievable, the final concentration would be fairly above today's level. A significant reduction of NO_x emissions has to be accounted for as well when considering alternative fuels.

Possible alternatives to today's transportation fuels are discussed. For road transportation a variety of alternative fuels exists which are available short term and can contribute to reduction of emissions, e.g. natural gas, biofuels and electric propulsion. For aircraft, energy content per mass is of primary importance, which eliminates many of the alternatives applicable to ground transportation.

In the long term, hydrogen is a promising alternative fuel for both air and ground transportation, with respect to both environmental compatibility and vehicle performance. For the weight sensitive aircraft, the liquefied condition is the only efficient way of storing hydrogen. For ground transportation, hydride and gaseous storage are feasible, but only liquid hydrogen (LH_2) offers vehicle performance comparable to today's gasoline cars.

The switch to an alternative, clean energy carrier has to be seen in the context of introducing clean, renewable energies in general. Technologies for renewable primary energies (hydropower, wave power, tidal energy, wind energy, solar and bio energy) required for the environmentally benign production of hydrogen exist, however, improvements in the economic use of those technologies is a long term prerequisite.

Regarding the application of alternative fuels, technologies for biofuels (alcohols, hydrocarbons), natural gas and electric propulsion are already state-of-the-art in ground transportation, some examples being:

- widespread use of biofuels in Brazil,
- about 800,000 cars worldwide running on natural gas,
- electric cars developed by numerous automobile manufacturers (usually limited to city traffic due to comparably low vehicle performances, resulting from today's battery characteristics).

The major application of hydrogen today is related to the space industry where it has been used as a propellant in the liquefied condition for 30 years. The technology for the liquefaction of hydrogen exists at an industrial scale, however, R&TD activities are necessary to improve the efficiency of the process. The experiences gained so far in LH_2 handling and application demonstrate that the hazard potential is not greater than that of conventional hydrocarbon fuels.

In contrast to space applications, the use of hydrogen in air and ground traffic up to now has been limited to studies (especially in aircraft design), experimental vehicles and demonstration projects. Currently, numerous automobile manufacturers are pursuing further development of hydrogen cars, e.g. BMW, Mercedes Benz AG, Mazda, Renault, Volvo etc. However, commencement of a series production, at present, is not seriously considered.

A project to introduce LH_2 in commercial aviation has been started by Deutsche Aerospace Airbus in 1988 and has since become a German-Russian cooperation, involving some 15 partner companies. The project considers topics like aircraft configurations and fuel system layout as well as infrastructural, safety, economical and ecological aspects.

A comprehensive hydrogen demonstration project is currently in progress in the scope of the Euro-Québec hydrogen project, covering numerous aspects of hydrogen production from clean energy sources, storage and distribution up to the end-use of hydrogen as a transportation fuel and in other applications.

List of Authors

Bakan, S.
Max-Planck-Institute of Meteorology
Bundesstr. 55
D-20146 Hamburg

Energy and the Environment

Borisov, V.
Tupolev Design Bureau
Naberejnaia Akademika
Russia-111250 Moscow

Experience from the TU-155 Laboratory Aircraft

Buchner, H.
Daimler-Benz Central Research
P.O. Box 800230
D-70567 Stuttgart

Characteristics of Alternative Fuels;
Technology for Gaseous Hydrogen Production;
Alternative Fuels in Ground Transportation

Grafwallner, F.
Deutsche Aerospace Space Systems
P.O. Box 801189
D-81663 München

Experience from LH_2 Space Application

Graßl, H.
Max-Planck Institute of Meteorology
Bundesstr. 55
D-20146 Hamburg

Energy and the Environment

Gretz, J.
Joint Research Centre
I-21020 Ispra (VA)

The Euro-Québec hydro-hydrogen pilot project

Klug, H.G.
Deutsche Aerospace Airbus
Kreetslag 10
D-21129 Hamburg

Characteristics of Alternative Fuels

LIST OF AUTHORS

Luger, P.
Deutsche Aerospace Space Systems
P.O. Box 801189
D-81663 München

Experience from LH$_2$ Space Application

Malyshev, V.
Tupolev Design Bureau
Naberejnaia Akademika
Russia-111250 Moscow

Experience from the TU-155 Laboratory Aircraft

Martin, D.J.
ETSU Strategic Studies Dept.
UK-Harwell, Didcot,
Oxfordshire OX11 0RA

Long Term Trends in Energy Consumption and in Availability and Cost of Conventional Fuel

Moon, D.P.
ETSU Strategic Studies Dept.
UK-Harwell, Didcot,
Oxfordshire OX11 0RA

Long Term Trends in Energy Consumption and in Availability and Cost of Conventional Fuel

Müller, M.
Deutsche Aerospace Space Systems
P.O. Box 801189
D-81663 München

Experience from LH$_2$ Space Application

Orlov, V.
SSSPE TRUD
Russia-443026 Samara

Aeroengines for Alternative Fuels

Pelloux-Gervais, P.
L'Air Liquide
Division Technique Avancées
B.P. 15
F-38360 Sassenage

Technology for Cryofuel Production and Handling

Pohl, H.W.
Deutsche Aerospace Airbus
Kreetslag 10
D-21129 Hamburg

Alternative Fuels in Aviation

Shengardt, A.
Tupolev Design Bureau
Naberejnaia Akademika
Russia-111250 Moscow

Experience from the TU-155 Laboratory Aircraft

Sulimenkov, V.
Tupolev Design Bureau
Naberejnaia Akademika
Russia-111250 Moscow

Experience from the TU-155 Laboratory Aircraft

Walther, R.
Deutsche Aerospace MTU
Dachauer Str. 665
D-80976 München

Aeroengines for Alternative Fuels

Wurster, R.
Ludwig-Bölkow-Systemtechnik
Daimlerstr. 15
D-85521 Ottobrunn

Clean Energy Sources

Hydrogen and other Alternative Fuels for Air and Ground Transportation

ABSTRACT

The proceedings compiled in this publication have been presented during a workshop which took place in June, 1993, in Brussels. The purpose of this workshop, involving some 40 experts from different industrial and research sectors, was to discuss the rationale and the possibilities for alternative fuels in air and ground transportation.

The proceedings summarize the state of knowledge regarding long term availability of conventional transportation fuels, based on fossil resources, and the environmental effects of fossil fuel use. Possible alternative fuels which can contribute to reduction of atmospheric pollution are discussed. In the context of a general switch to clean, renewable energy sources, hydrogen is seen as a perspective energy carrier for both air and ground transport.

Application of alternative fuels in transportation up to now has been limited to experimental vehicles and demonstration projects, e.g. cars operating on natural gas, electric cars for city traffic, experimental cars using hydrogen, experimental aircraft TU-155 operating on liquid hydrogen or liquid natural gas. Current projects aiming to disseminate the use and to improve the commercial competitiveness of alternative fuels are outlined. As an example of a comprehensive demonstration project in the field of alternative fuels, a survey of the Euro-Québec hydrogen project, covering numerous aspects of hydrogen production from clean energy sources, storage and distribution up to the end-use of hydrogen as a transportation fuel and in other applications, is presented.

1
Long-term Trends in Energy Consumption and in Availability and Cost of Conventional Fuel

D.J. Martin, D.P. Moon
ETSU, Harwell, UK

1.1 INTRODUCTION

This chapter reviews trends in global energy requirements over the next 40–60 years, and in the fuel requirements for air and road transportation. The price trends and availability of conventional fuels are identified, and prospects for alternative fuels are presented. Comparison between fuels, under a range of scenarios, indicates the timescale within which alternative fuels could become price-competitive in transport applications. A particular focus is placed on the introduction of hydrogen as an energy source.

1.2 TRENDS IN GLOBAL ENERGY REQUIREMENTS

Long-term trends in energy requirements depend on the rates of growth in population and in per capita energy demand. Per capita demand is itself dependent on:

- the rate of economic growth (in a given country);
- changes in the energy intensity of economic activity (i.e. total primary energy requirements per unit of real GDP);

- the price signals provided by the supply-demand balance on world energy markets or by government actions (for instance, in response to environmental concerns).

The world population in 1990 was about 5.3 billion [1]. It has been doubled over the past 40 years, and might be expected to double again by 2050 [2].

Population growth is higher in developing countries than in industrialized countries (see Table 1.1), therefore projections of global energy requirements will depend critically on what assumptions are made about future per capita energy demand in developing countries. This demand is likely to grow rapidly because of:

- rapid growth in economic activity and in standards of living;
- increasing industrialization and transport usage;
- diminishing availability of non-commercial sources of energy (such as biomass), partly due to increased urbanization.

In addition, limitations on financial resources are likely to reduce the ability of developing countries to take maximum advantage of energy technologies that could enhance efficiency and inter-fuel substitution.

World energy demand (for commercial energy) has increased by about 40% since 1973 [3]. Per capita demand in developing countries has increased by more than 50% since that date, and total demand in this group has more than doubled. In contrast, per capita energy demand in the OECD was almost the same in 1988 as in 1973, while energy intensity fell at an average annual rate of about 1.9%. CIS and East European countries increased their per capita demand over this period.

It is a common belief [2, 4] that the average of zero growth in per capita energy demand in the OECD since 1973 will be continued in the near future, as a result of saturation of demand, continued investments in energy conservation and efficiency, and structural transformation of OECD economies towards less energy-intensive sectors. However, there is less common

Table 1.1 Population projections to 2050 (Source: Eden [2])

	Population (billion)	
	1988	2050
OECD	0.77	0.88
CIS/East Europe	0.42	0.48
Developing Countries	3.92	9.16
World	5.11	10.52

ground on longer-term forecasts. The EC 'Conventional Wisdom' reference scenario shows per capita energy demand increasing by nearly 25% between 1987 and 2010, although a decline is forecast under its efficiency scenarios (which assume specific policy initiatives) [5]. For the CIS and East European countries, the present inefficient use of energy accompanies general economic inefficiency, therefore it might be assumed that longer-term economic growth could be achieved without a substantial increase in per capita energy demand.

Eden [2] projects two broad scenarios for future energy requirements:

- the targeted growth (TG) scenario assumes that per capita energy demand in the industrialized countries (OECD and CIS/ East Europe) as a whole is the same in 2050 as in 1988, namely about five toe per capita (one toe is a tonne of oil equivalent, or 42 GJ). This is stated to be probably within the plausible range of many energy analysts. Per capita demand in the developing countries is assumed to grow threefold, from 0.5 toe average to 1.5 toe; by comparison, a continuation of the average growth in per capita demand during 1973–88 would lead to a sixfold increase by 2050 for these countries;

- the targeted efficiency (TE) scenario assumes that per capita energy demand in the industrialized countries is halved by 2050. This is stated to be probably near the extreme lower end of any range that would be accepted by most energy analysts. Average per capita energy demand in the developing countries is assumed to double by 2050. This scenario is based on circumstances of extreme environmental concern, leading to government actions to reduce energy consumption and CO_2 emissions.

The energy demand implications of the TG and TE scenarios are shown in Tables 1.2 and 1.3. For targeted growth, world energy demand increases from 7.9 Gtoe in 1988 to more than 20 Gtoe in 2050. The average growth in world energy is 1.6% pa (which may be compared with 2% pa from 1973 to 1990). This could be associated with economic growth per capita of 2% pa

Table 1.2 Targeted growth (TG) scenario energy demand (Source: Eden [2])

	Per capita energy demand (toe)		Total demand (Gtoe)	
	1988	2050	1988	2050
OECD	5.2	5.2	4.0	4.6
CIS/East Europe	4.4	4.4	1.9	2.1
Developing Countries	0.5	1.5	2.0	13.8
World	1.5	2.0	7.9	20.5

Table 1.3 Targeted efficiency (TE) scenario energy demand (Source: Eden [2])

	Per capita energy demand (toe)		Total demand (Gtoe)	
	1988	2050	1988	2050
OECD	5.2	2.6	4.0	2.3
CIS/East Europe	4.4	2.2	1.9	1.1
Developing Countries	0.5	1.0	2.0	9.2
World	1.5	1.2	7.9	12.6

worldwide, assuming that economic growth exceeds energy growth by 2% pa in the industrialized countries (in line with OECD performance since 1979).

The targeted efficiency scenario leads to world energy demand in 2050 equivalent to 12.6 Gtoe. This would require an additional 1% pa better improvement in energy intensities than with targeted growth, in industrialized countries.

By comparison, IIASA predict a 1.4% average annual growth rate in global primary energy consumption over the period 1990–2020, as the 'conventional wisdom' of the international energy community [6]. This is based on a compilation of experts' projections for middle-of-the-road, business-as-usual, surprise-free scenarios. The IEA predicts 2.5% pa growth in world energy consumption from 1987 to 2005 under a rising oil price scenario, and slightly higher growth under a constant oil price scenario [4]. The EC/DGXVII predicts 2.0% pa growth from 1987–2010 in its 'conventional wisdom' scenario [5].

1.3 TRENDS IN FUEL REQUIREMENTS FOR AIR AND ROAD TRANSPORTATION

The level of demand for transport and the mode of transport required in the future depend on a large number of factors including geographic, demographic, economic, environmental, technical and policy related. Together, these underpin the desire for mobility. However, the satisfaction of transport requirements is also dependent on how government responds to changes in these factors, both directly with regard to its transport policy, and indirectly with regard to land use policies, regional policies and even industrial and employment policies.

Economic activity is one of the key overall determinants of the level of demand for travel. At constant relative prices and with sufficient capacity available, there is a tendency for the amount of transport to grow rather more than in proportion to the level of economic activity.

A particular characteristic of road and air transport is the technical difficulty in switching fuels, by comparison with other energy applications (such as electricity generation and space heating). This results from the need for high energy density fuel sources. As a result, liquid hydrocarbons are likely to be reserved preferentially for transport use in the future; as an example, the IEA's rising oil price scenario predicts the rise in OECD oil demand to continue to come mainly from road and air transport and from the petrochemical industry [4].

The EC/DGXVII has studied transport demand under four different scenarios over the period 1990–2010 [5]:

- In scenario 1, the EU economy grows at 2.7% pa, and overall world economic growth averages 3% pa. Non-OPEC supply of oil is predicted to plateau, and higher world oil demand translates almost directly into an additional call on OPEC (allowing OPEC to exercise greater market power). Scenario 2 assumes stronger economic growth from 1990 to 2000. In both scenarios, transport is the fastest growing sector of the economy, increasing energy use by 20–35% despite improvements in vehicle efficiency and rising fuel prices. Oil continues to dominate the transport sector.

- In scenario 3, strong economic growth (3–3.5% in the EC, 3.5–4% worldwide) stimulates investment in more efficient technology, helped by policy initiatives aimed at reducing energy intensity. Scenario 4 assumes more moderate economic growth (similar to scenario 1), and higher end-user prices (including carbon taxes). In transport, there is a major shift in demand towards much greater use of public transport; under scenario 4, transport fuel consumption declines by over 37% from 1987 to 2010 (2% pa overall).

The impact of these scenarios on ground transport is summarized in Table 1.4. Scenarios 3 and 4 show a substantial switch away from the use of private cars towards railways and public transport. However, air transport continues to expand under all scenarios (by 2.8% pa in scenario 1, 3.9% in scenario 3); growth in demand for jet fuel slows down as the size of planes and load factors rise, and fuel efficiency improves.

The LOTOS Panel [7] confirms this prediction of growth in air traffic, predicting a global long-term average rate of growth of passenger traffic demand of about 4.5% pa and of freight traffic of about 8% pa, almost

6 LONG-TERM TRENDS FOR CONVENTIONAL FUELS

Table 1.4 EC/DGXVII Scenarios for the transport sector up to 2010 (Source: EC/DGXVII [5], modified)

Scenario 1 (Conventional Wisdom)	Scenario 2 (Driving into Tensions)
Increasing road traffic Increasing air traffic Efficiency improvements	Road congestion, due to • rapid economic growth • lack of traffic management policies Deterioration of road fuel efficiency Increasing air traffic
Higher fuel prices (carbon tax), leading to slowdown of transport activities Expansion of railways and public transport Energy saving measures Increasing air traffic	Decrease of road traffic Expanison of railways and public transport Promotion of most efficient vehicles Increasing air traffic
(High Prices) Scenario 4	(Sustaining High Economic Growth) Scenario 3

trebling air transport demand in the next 25 years. This projection assumes average economic growth, worldwide, of 2.7–2.9% pa. It also assumes that shortages in the essential elements of air transport supply infrastructure do not become limiting; this will require expansion of airport capacity, and major technical and organizational changes in air traffic management.

The level of air transport demand can be related to the buoyancy of economic activity. Historical data show business travel to be largely dependent on GDP and volumes of business, while leisure travel is linked to the disposable income of individuals [8]. Fuel prices are important in the short run, as witnessed by the effect that abrupt changes in fuel price have had in the past. Domestic and long haul air travel demands may be assumed to be most affected by higher fuel prices. In the case of domestic air travel, there are greater opportunities for substitution by other transport modes. In the case of long haul travel, the fuel price effect is greater since fuel costs comprise a greater proportion of the cost of flying.

1.4 ENERGY RESERVES AND RESOURCES

It is estimated by Rogner et al. [9] that global fossil energy reserves will last for about 130 years at the 1990 level of total global energy consumption (see Table 1.5). Total resources and possible occurrences could last several times as long. Eden [2] has projected the future rate of depletion of reserves, based on assumed upper limits on production rate, as summarized in the following paragraphs.

Oil is likely to be available throughout the next century, with demand varying beneath a supply ceiling. This ceiling will probably remain below 4 Gtoe pa, and decline slowly from about year 2030. Investment in high-cost oil production will continue to be inhibited by the potential availability of flexible but volatile production from the Middle East – limited by policies or crises, rather than by resource availability or cost.

The Middle East holds two-thirds of the world's proved reserves of oil. These would last for about 90 years at 1 Gtoe pa average production. Elsewhere, production currently totals over 2.2 Gtoe, but the reserve/production ratio is only 21 years. The use of improved production technology (notably in the CIS), coupled with further discoveries, could maintain output near to 2 Gtoe for several decades. At that stage, conventional oil could be supplemented by more extensive development of heavy oil deposits, tar sands and shale oil. Total reserves and resources of unconventional oil are much larger than those of conventional oil (Table 1.5), and they could help to maintain total world oil production near to 3 Gtoe throughout the 21st century.

Identified reserves of natural gas are equivalent to 110 Gtoe, and it is likely that the ultimate resources of conventional gas are comparable with those of oil (Table 1.5). This would allow production of 3 Gtoe to be maintained during the 21st century; current demand is 1.7 Gtoe pa, and is likely to increase steadily as more natural gas is used to substitute for oil and coal, particularly for electricity generation. The use of natural gas in competition with oil should help to stabilize world oil demand and to moderate oil price rises.

Hydropower could continue to increase, particularly in developing countries which need to reduce the cost of energy imports. However, its growth will slow as options become more costly and/or more distant from major markets, and as environmental concerns over land use are given higher priority. The world total could rise to the equivalent of 1 Gtoe pa, twice its current level, by the year 2050 [2]; other estimates indicate a lower potential contribution (0.6 Gtoe in 2030) [10]. Additional hydropower investment might be stimulated if demand for hydrogen as an alternative fuel increased rapidly at some future date; electrolysis of water using hydropower allows low-cost hydrogen production, while avoiding the energy losses associated

8 LONG-TERM TRENDS FOR CONVENTIONAL FUELS

Table 1.5 Global fossil energy reserves, resources and occurrences (TWyr, Source: Rogner [9])

	Consumption 1990	Reserves		Total reserves plus resources	Additional occurrences
		Identified	Remaining to be discovered		
Oil – conventional	4.4	194	79	273	> 320
– unconventional	–	240	–	525	> 4600
Gas – conventional	2.5	144	136	280	> 320
– unconventional	–	–	–	635	> 700
– clathrates	–	–	–	–	20,000
Coal	3.1	701	–	3880	> 4100
Total	10.0	1279	> 215	> 5590	> 30,000

with long-distance transmission of electricity from remote hydropower locations.

The remaining commercial energy supply to meet future world demand will involve coal, nuclear power and new renewable sources (geothermal, wind, solar, oceans, biomass and energy crops). The size of the contribution from these sources can be regarded as uncertain – it will depend on:

- the level of energy demand;
- future costs;
- technological developments;
- social concerns over environmental impact;
- national policies towards the exploitation of domestic resources.

Coal production could be expanded to 8 Gtoe pa by 2050 (nearly four times current levels), increasing at about 2.7% pa, well within historical rates of growth. However, such rates of growth in the past have generally been associated mainly with indigenous demand in the major producing countries. If transport and trading problems are taken into account, a figure of 5 Gtoe would be more likely [2]. In addition, coal is regarded as a 'dirty' fuel in terms of CO_2, NO_x, SO_2 and other pollutant emissions (per unit of energy delivered), and therefore is likely to be targeted by any environmental constraints.

Nuclear production is currently equivalent to 0.45 Gtoe pa. If nuclear power became a high priority within the next two decades (e.g. in response to heightened environmental concern over global warming), capacity could increase to 2–3 Gtoe by 2050, particularly in industrialized countries with adequate financial resources and know-how. However, currently there is a lack of consensus at both the political and public opinion levels about nuclear's role in the future. This is not only due to safety concerns. There is a mismatch in many countries between the financial requirements of utilities, which need the flexibility of short lead times and low unit capital costs, and the financial realities and risk concentration of nuclear power. Thus nuclear production could fall quite rapidly after the year 2000 as present capacity is retired, and its contribution to world energy by 2050 could be less than 0.1 Gtoe.

The potential from new renewable sources of energy is uncertain, on technical and economic grounds. Eden [2] suggests an upper limit of 3 Gtoe pa, but notes that this may not be achievable since the capital investment required for new renewables is likely to be comparable with that for nuclear power. Other estimates indicate a total potential of 5 Gtoe of primary energy (equivalent to 1.7 Gtoe electrical energy, at 34% conversion efficiency) by 2030 [10].

1.5 PRICE TRENDS FOR CONVENTIONAL FUELS

Price forecasts for oil have achieved a very poor record of accuracy. The main reason for this is that oil prices do not typically depend on any one behaviourally predictable factor. The fundamental rules of supply and demand can identify price boundaries, but within this range oil prices are dependent on a variety of economic, political and technological factors. In the long term these uncertainties include:

- how the CIS economies will develop their fossil fuel resource (for export or for domestic use);
- whether growth in developing countries will increase oil demand or be fuelled by an alternative;
- how the political stability and market power of Middle East countries will develop;
- how environmental policies and transport demand will impact on oil demand.

The price floor for oil can be defined as the minimum price at which exploration will be financed and reserves will continue to be acquired; although the price may fall below this level in the short term, it would not be sustainable. This price is probably around $17/barrel (1990 money), sufficient to maintain exploration and development activity both in the Middle East and in many non-OPEC regions [8].

The price ceiling for oil can be defined as the level at which expensive unconventional oil production technologies like shale oil, tar sands or coal liquefaction become economically attractive. A sustainable price ceiling is considered to be around $50/barrel [8].

As in the case of oil, the prospects for internationally traded coal are influenced by a variety of economic and political factors on both the supply and the demand side. Historically, since 1975, the coal price appears to track the spot crude oil price, although the significance of this is not clear; any correlation may be largely coincidental. The major uncertainties concerning future prospects for coal are on the demand side, and relate to:

- the need for coal to bridge an uncertain gap between growing world energy demand, especially from developing countries, and the feasible supply from other energy sources;
- environmental legislation to limit emissions that contribute to both acid rain and global warming.

The single most important factor influencing the future of the gas market is the increase in demand for gas for electricity generation, where the new generation of combined cycle gas turbines offers many advantages over conventional generating technology. If the principle of pricing according to market value continues to prevail, then the large quantities of gas required for electricity generation will have to be priced to compete with coal-fired generation. This could imply a relative reduction in price which may not guarantee the availability of supplies. It is possible that the pricing convention will have to change towards a valuation based on production costs.

At present the market for uranium is characterized by soft prices and a substantial surplus. Under currently forecast levels of world demand, uranium is likely to remain in surplus until well into the next century.

The degree of upward pressure on energy prices can be inferred by examining different scenarios for world energy demand; Eden's examples, described previously, provide a broad framework for analysis. His targeted growth projection requires an increase in the world energy supply from 7.9 Gtoe in 1988 to 20.5 Gtoe in the year 2050. An illustrative possibility for energy production to meet this need is given in Table 1.6. Oil and gas provide 6 Gtoe, hydro provides 1 Gtoe, and new renewables provide a maximum of 3 Gtoe. This leaves 10.5 Gtoe to be provided by coal and nuclear, currently totalling 2.6 Gtoe. This is roughly equal to the maximum feasible supply from these sources, and indicates that severe energy stresses and price pressures would result from a TG future. Such a future would also involve large increases in carbon emissions and in nuclear capacity, which could prove unacceptable socially and politically, exacerbating the demand-supply imbalance. Constant per capita energy use in industrialized countries, coupled with their continuing economic growth, could provide a willingness and ability to pay higher prices for convenient sources of energy, notably oil and gas.

Table 1.6 Energy supply in a targeted growth scenario (Source: Eden [2])

	World energy (Gtoe)	
	1988	2050
Oil	3.1	3.0
Gas	1.7	3.0
Hydro	0.5	1.0
New renewables	–	3.0
Coal	2.2	8.0
Nuclear	0.4	2.5
Total	7.9	20.5

Table 1.7 Energy supply in a targeted efficiency scenario (Source: Eden [2])

	World energy (Gtoe)	
	1988	2050
Oil	3.1	3.0
Gas	1.7	3.0
Hydro	0.5	1.0
New renewables	–	0.5–1.5
Coal	2.2	3.0–5.0
Nuclear	0.4	0.1–1.5
Total	7.9	12.6

In the targeted efficiency scenario, 7 Gtoe could be provided by oil, gas and hydro (Table 1.7). The remaining need for 5.6 Gtoe could be provided entirely by increased coal production (with a consequent doubling of carbon emissions). Even with reluctance to use coal for environmental reasons, it is unlikely that coal production would be less than 3 Gtoe, since coal is a major resource for future growth in key developing countries (notably India and China). It is also likely to remain an important component of energy production in the USA and the CIS, partly to improve security of supply and reduce energy import costs. Energy prices under this scenario may be controlled more by government actions (such as the imposition of energy and carbon taxes), designed to limit per capita energy use, than by market forces.

1.6 AVAILABILITY OF ALTERNATIVE FUELS

Estimates of the potential global resources of alternative fuels are restricted here to those non-fossil fuel energy sources, and their derivative fuels (such as hydrogen), which are thought most likely to be significant global scale resources in the period up to 2040 (excluding nuclear power, discussed previously). It should be noted that much of the literature on this subject draws attention to the potential size of the resource based on technical considerations, while the actual resource that could credibly be made available will depend on complex social, environmental and political factors as well as on economic comparisons. For instance, many of the renewable options make considerable demands on land use, with consequent environmental and ecological impacts.

Swisher and Wilson [10] have recently reviewed the literature on renewable energy potentials. Their estimates are summarized in Table 1.8. For hydropower, the main technical problem is energy transmission to the principal centres of industry and population. Conversion to hydrogen is likely to be favoured at remote locations developed for export purposes. However, the cost of much of the future hydropower capacity will be well above the low cost levels of much of the current capacity. By the year 2000, the global average cost of hydropower at source could be £11–14/GJ [11].

For solar power, capital costs and device efficiency are the major sources of uncertainty in predictions. Ogden and Williams claim that to produce hydrogen equivalent to US oil use would require a solar collector field equivalent to about 7% of its desert areas [12]. Future power costs would be about £3–6/GJ.

In principle, wind power can supply a large fraction, possibly all, of global energy requirements [13]. The practical problem is to develop wind power technologies which are sufficiently cheap and reliable to produce reasonably priced electricity. For UK conditions, estimated electricity prices from wind range from about 5/GJ to 36/GJ as the cumulative electricity generated rises from 0.1 to 48 TWh/y with the use of progressively less favourable sites [14].

Biomass is already a major energy resource, accounting for 13% of the world's annual energy use of 335 EJ/yr, and for 43% of the energy use of the Third World, mainly as fuel wood by rural populations [15]. The use of agricultural land for food production and forestry for fuel wood will continue to take priority, especially in poorer countries of the world. However, there will be biomass available from forestry and agricultural residues and from switching to growing energy crops rather than producing surplus food in some areas, which could be used as a commercial energy source. Hall and de Groot [15] conclude that 5–10% of Europe's energy requirements could be met by biomass by the year 2000, and quote an IIASA study as estimating that the requirement for road transport fuel in Western Europe could be met from biomass grown on 5% of the land area. This would

Table 1.8 Estimates of practical potentials of renewable energy sources in 2030 (Source: Swisher and Wilson [10])

	Energy resource (TWh$_e$/yr)
Hydropower	7100
Geothermal	1500
Wind	4900
Solar	1400
Oceans	250
Biomas	8000
Energy crops and plantations	3600

Table 1.9 Estimates of production costs for hydrogen (Source: ETSU [8])

Production method	Production cost (£/GJ, 1990 values)
Electrolysis of water, using solid polymer electrolytes	
– UK current generating mix	18.6
– hydropower	8.5
Gas reforming	3.6
Coal gasification	9.3
Biomass gasification	8.0

represent a substantial switch in land use, and therefore its feasibility must be questioned.

The use made of biomass will depend on local needs. Possible applications include:

- electricity generation;
- conversion to liquid fuels, such as ethanol and methanol;
- conversion to hydrogen.

The conversion of woody residues to methanol can yield fuels at prices competitive with gasoline if the costs of biomass are held to the lower end of the present range [13]. Also, biomass gasification may provide a low cost route to hydrogen production. Nevertheless, policy decisions and competing demands for land use are likely to be key determinants of biomass availability as a commercial energy source.

There are four basic methods for the production of hydrogen: electrolysis of water, reforming of natural gas, gasification of coal, and gasification of biomass. Comparative costs are given in Table 1.9. Currently, gas reforming is the lowest cost route; however, the future choice of production method would depend on relative movements in world prices for the different energy inputs (electricity, gas, coal, biomass), and on the pressure to switch to non-fossil inputs in response to environmental concerns.

1.7 ADOPTION OF ALTERNATIVE FUELS FOR AIR AND ROAD TRANSPORT

Energy use in air and road transport is currently dominated by conventional oil supplies. The preceding analysis has indicated that oil will continue to be available throughout the 21st century, with a gradual phasing-in of unconventional sources. Therefore, the introduction of alternative transport fuels for economic reasons is not simply a supply-side issue – it will depend on a complex interaction of factors such as:

- trends in transport demand;
- relative movements in conventional fuel prices;
- the ease of substitution for oil in other sectors of the economy;
- trends in technical development, hence cost reduction, for alternative fuel technologies;
- trends in technical development, hence efficiency savings, for transport technologies;
- external factors such as oil supply shocks and environmental concerns.

This interplay has been explored in a set of scenarios developed by ETSU [8]. These scenarios have provided the basis for energy systems modelling to yield projections of future UK energy supply and demand (over the period 1990–2030). The outcomes relevant to air and road transport are summarized below.

The alternative fuels considered for road vehicles were compressed natural gas, biomass-derived ethanol, methanol, liquid hydrogen and various substitute gasoline and diesel fuels (such as biodiesel, and coal and natural gas derived gasoline and diesel). In addition, electric and hybrid vehicles were included in the model. Energy use by vehicles was assumed to improve steadily due to changes in vehicle design.

A comparison of total costs (fuel and non-fuel) between different vehicle types showed that, for small cars, there is a possibility that compressed natural gas, direct injection diesel, leanburn gasoline and battery powered vehicles will be required in the UK in the next 40 years. No alternative technology is attractive at any point in the large car market. Ethanol, liquid hydrogen and fuel cell power are unlikely to become attractive in private car transport at any point during the period of the scenarios. Synthetic gasoline and diesel produced by coal liquefaction, and biomethanol and biodiesel, are feasible long-term options as substitute transport fuels (e.g. first contributing in 2025 in scenarios of heightened environmental concern).

The additional weight, volume and cost incurred by present hydrogen vehicle technologies is likely to restrict their near-term application to commercially operated fleets such as urban buses and vans. The advantages of hydrogen in this niche application are:

- the relative ease and low cost of setting up the distribution and supply infrastructure, based on central refuelling depots;

- the relative ease of incorporating hydrogen storage equipment into large vehicles.

However, the scenario projections indicate that CNG fuelled vehicles are more competitive than hydrogen vehicles at the point where a switch from conventional petrol and diesel is required (due to rising oil prices). Hydrogen as a road fuel could only become a viable option in the following circumstances:

- in the medium term if there is stringent control of vehicle emissions – under tight CO_2 constraints corresponding to the IPCC-D scenario, hydrogen-fuelled buses start to be selected around 2025;
- in the medium term if there is utilization of regional energy niches (e.g. for solar and hydropower) to produce hydrogen at low cost;
- in the long term when a substitute for fossil fuels is required due to exhaustion of reserves (probably not earlier than the 22nd century);
- in transportation tasks in special sectors which have extremely stringent requirements on environmental compatibility.

For air transport, the ETSU modelling projected that liquid hydrogen (LH_2) may become economically competitive as early as 2010 in scenarios of heightened environmental concern. 2010 is also projected to be the earliest date of availability of LH_2 fuelled aircraft. The major application of hydrogen would be in long-haul international flights, where fuel costs form a larger proportion of the total costs. LH_2 is additionally adopted in 2025 in the event of high oil prices; in other scenarios, LH_2 is not cost-competitive within the time horizon considered.

The least-cost route for UK production of LH_2 after the year 2010 is projected to be gasification of energy crops (short-rotation forestry), under scenarios of heightened environmental concern or high oil prices. Most of the hydrogen produced is liquefied for use in aircraft. However, some is used by 2025 in gaseous form to mix with natural gas for domestic and commercial heating. LH_2 could also be imported to the UK from countries with a large resource of lowcost renewable energy; again, such an option would be more favoured under future conditions of heightened environmental concern.

1.8 SUMMARY

Projections based on current trends in the rates of growth of population and per capita energy demand indicate that by the year 2050 world energy supplies will only just meet demand, and CO_2 emissions will be much increased. Moreover, these supplies will have to include greatly expanded production of coal, nuclear power and renewable energy. Such expansion might not be feasible, owing to environmental concerns, socio-political objections, and problems with capital investment and technological development. Energy prices would rise sharply if a significant demand-supply imbalance developed.

If governments take action to improve energy efficiency and reduce CO_2 emissions, in response to environmental concern, then the stress on energy supplies will be reduced. End-user energy prices could be increased by carbon/energy taxes. This would provide an economic incentive for switching to alternative fuels, such as hydrogen produced from biomass or renewables. Other government regulations might mandate such changes.

Demand for road and air transport is forecast to grow substantially as economic growth continues. This will increase the demand for oil as the dominant energy source for transport, despite rising oil prices and efficiency savings, unless governments adopt specific measures which promote a switch to public transport. Air traffic is projected to increase under a range of scenarios.

Substitution for oil as a transport fuel is unlikely to be required during the 21st century, for reasons of supply exhaustion alone, since total reserves and resouces of conventional and unconventional oil are substantial. The introduction of alternative fuels will instead depend on their price competitiveness and on government intervention. Feasible fuels include compressed natural gas, electric power, synthetic gasoline and diesel, biomethanol and biodiesel, and hydrogen.

The timescale for fuel switching is scenario dependent: heightened environmental concern, and high oil prices (e.g. due to OPEC domination of oil markets or an oil supply shock resulting from political disturbances in major exporting countries), are projected to encourage early adoption of alternative fuels.

Hydrogen could be adopted as early as 2010 in long-haul air transport, if governments react in a radical way to environmental concerns. Alternatively, hydrogen could become competitive around 2025-2030 as a result of high oil prices; at this point, oil prices may be influenced more by the rising energy demand in developing countries, than by political factors in supplier countries.

Hydrogen uptake in road vehicles is more uncertain than for aircraft, since more alternatives to petroleum fuels exist. In the absence of specific government actions to mandate the use of hydrogen, niche applications (such as bus fleets) may become economically viable in 30-40 years time. This would depend on the degree of environmental concern and the rates of technical progress in vehicle design for different fuel types. It would also depend on the rates of change in technologies for renewable energy, hydrogen production and hydrogen distribution, affecting fuel costs. Countries with lowcost production routes for hydrogen (e.g. from renewables or biomass) could adopt this as a transport fuel at an earlier stage.

1.9 REFERENCES

1. United Nations, *'Demographic Yearbook'* (annual), UN, New York.
2. R.J. Eden, World energy to 2050, *Energy Policy*, Volume 21, 1993, pp. 231–237.
3. British Petroleum, *Statistical Review of World Energy*, BP, London, 1990.
4. OECD/IEA, *Greenhouse Gas Emissions: The Energy Dimension*, OECD/IEA, Paris, 1991.
5. Conference papers and report on *Major Themes in Energy Revisited, Energy for a New Century: The European Perspective*, EC/DGXVII, Brussels, May 1990.
6. N. Nakicenovic et al., Long-term Strategies for Mitigating Global Warming, *Energy*, Volume 18, 1993, pp. 403–419.
7. Panel on Long-term Outlook for Aeronautics Research and Technology (LOTOS), *Flying Ahead: A View of the Future for Civil Aeronautics*, EC/DGXII, Report EUR-13334 EN, January 1991.
8. Energy Technology Support Unit (ETSU), Unpublished information.
9. Rogner, N. Nakicenovic and A. Grubler, Second and Third Generation Energy Technologies, *Energy*, Volume 18, 1993, pp. 461–484.
10. J. Swisher and D. Wilson, Renewable Energy Potentials, *Energy*, Volume 18, 1993, pp. 437–459.
11. H.K. Schneider and W. Schultz, Investment Requirements of the World Energy Industries, *World Energy Conference*, 1987.
12. J.M. Ogden and R.H. Williams, *Solar Hydrogen: Moving beyond Fossil Fuels*, World Resources Institute, USA, 1989.
13. Energy Technology Support Unit (ETSU) and University of East Anglia, Unpublished Information.
14. *Energy Paper* No. 58, UK Government, 1989.
15. D.O. Hall and P.J. de Groot, *Biomass*, John Wiley and Sons, 1987.
16. G.R. Davis, Energy for the planet earth, *Scientific American*, Volume 263, September 1990, pp. 21–27.

2
Energy and the Environment

H. Graßl, S. Bakan
Max-Planck Institute of Meteorology, Hamburg, Germany

2.1 INTRODUCTION

Todays global primary energy production is mainly (to about 85%) based on the combustion of fossil fuels. These are widely available, simple to handle and use, and rather cheap. However, it has become evident in recent years, that this energy production results in severe environmental problems. The most prominent of these on the global scale has become the anthropogenic greenhouse effect due to CO_2 emissions. Also the production of tropospheric ozone as a consequence of the release of NO_x is a major point of concern, since ozone is an even more effective greenhouse gas than CO_2. Additional mainly local or regional problems result from the emission of CO, these include unburnt hydrocarbon fractions, sulphur dioxide and soot particles.

The importance of the rapid increase in the atmosphere's greenhouse effect by the anthropogenic release of CO_2 (and some other gases, Fig. 2.1) has been realized only within the last decade. The overview study of the International Panel on Climate Change [1], a UN initiative to evaluate the greenhouse threat, summarized the generally accepted concern of scientists about the possible consequences. It is held that the average temperature of the Earth's atmosphere will increase by several degrees within the next century (Fig. 2.2) and that sea levels will rise by tens of centimetres if the carbon dioxide concentration in the Earth's atmosphere increases at the present rate.

As the present day CO_2 concentration is already more than 25% higher than the pre-industrial value, the observed global mean temperature increase (Fig. 2.3) during this century is a plausible consequence. However, present (or paleoclimatic) observations of climate indicators (like global mean temperature) do not yet represent a statistically significant proof of the envisioned consequences of an increased greenhouse effect. Nevertheless, the potentially disastrous consequences of the anthropogenic heating

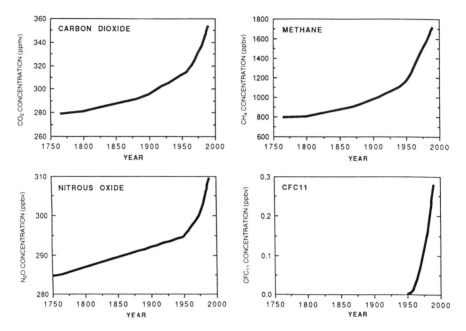

Fig. 2.1 Concentrations of carbon dioxide and methane, after remaining relatively constant up to the 18th century, have risen sharply since then due to man's activities. Concentrations of nitrous oxide have increased since the mid-18th century, especially in the last few decades. CFCs were not present in the atmosphere before the 1930s [1].

of the Earth's atmosphere caused 154 countries (including all industrialized states) to sign the Framework Convention on Climate during the United Nations Conference on Environment and Development [2] in Rio de Janeiro in June 1992. This convention calls for a stabilization of greenhouse gas concentrations in order to avoid massive climate change threatening ecosystems, food production, and economic development. In order to reach such a stabilization a reduction of global CO_2 emission by about 60% will be necessary (Fig. 2.4).

The use of fossil fuels for air and ground transportation contributes strongly to both transfrontier air pollution and greenhouse gas accumulation in the atmosphere. Nearly a quarter of total primary energy consumption is contributed by the traffic sector in the (former) Federal Republic of Germany, with still growing percentage (Fig. 2.5). More than 90% of this energy is produced from fossil fuels, mainly mineral oil. Since traffic thus contributes directly and considerably to two out of three global environmental problems associated with the atmosphere (increased greenhouse effect, changed tropospheric chemistry, ozone depletion), new fuels, less

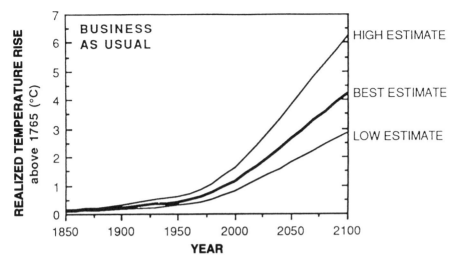

Fig. 2.2 Simulation of the increase in global mean temperature from 1859–1990 due to observed increases in greenhouse gases, and predictions of the rise between 1990 and 2100 [1]. Business as usual refers to a scenario with CO_2 emissions increasing in future at the same speed as in the past (see Fig. 2.4).

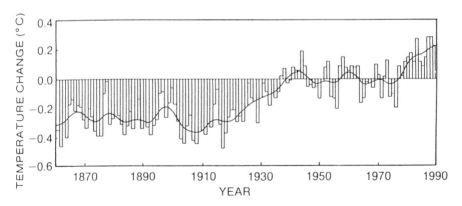

Fig. 2.3 Global-mean combined land-air and sea-surface temperatures, for the time period 1861–1989 relative to the average for 1951–1980 [1].

environmentally damaging, have to be sought now, even without any fossil fuel shortage. Since the anthropogenic greenhouse effect will soon – after the ratification of the UN Framework Convention on Climate – cause governments to adopt measures according to certain protocols, it is wise to act now.

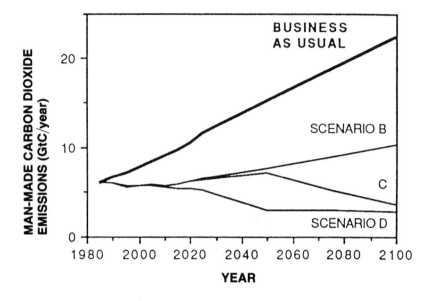

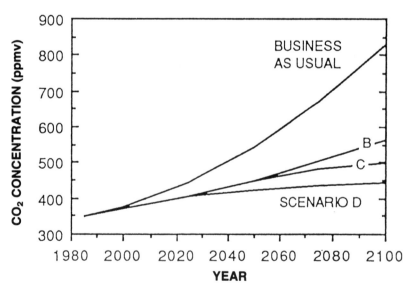

Fig. 2.4 Emission scenarios (upper panel) used by the IPCC [1] for the prediction of future CO_2 concentration development (lower panel). Only scenario D, which calls for a 60% CO_2 emission reduction by 2050, results in a stabilization of CO_2 concentration within the next century.

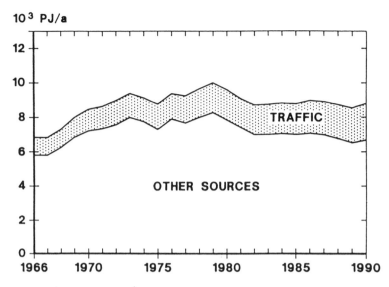

Fig. 2.5 Development of the contribution of the traffic sector to total energy consumption in Germany, without new federal states [3].

2.2 PROBLEMS FROM FOSSIL FUELS

The energy production by combustion of fossil fuel in the vehicles' motors causes the emission of a variety of exhaust products. Table 2.1 presents the amount of primary and the type of secondary combustion products for kerosene as an example for the emission from fossil fuel propellant systems. The different environmental consequences of traffic exhaust are discussed according to the lifetime of substances in the atmosphere.

2.2.1 Short-lived pollutants

The strong airflow in the atmosphere distributes all gaseous substances and particulates with less than a few micrometres diameter rapidly across borders, coastlines and mountain ranges. Therefore, even short-lived substances like nitrogen oxides NO_x (= NO and NO_2) with a lifetime between hours and a few days may influence large areas. A prominent example is the nearly continent-wide summertime photochemical smog with high ozone concentrations, caused mainly by emissions of NO_x plus a variety of partly burnt or unburnt hydrocarbons into a sunlit atmosphere. It is believed that the observed increase of tropospheric ozone concentration of about 2% per year (Fig. 2.6) is mainly due to this source. While the total amount of NO_x

Table 2.1 Combustion products of kerosene (as an example for a fossil fuel), liquid natural gas (LNG), and liquid hydrogen (LH_2) according to Klug and Graßl [4]

Fuel	Kerosene	LNG (Methane)	LH_2
Mass for equal energy content	1 kg	0.856 kg	0.357 kg
Primary combustion products	3.16 kg CO_2	2.35 kg CO_2	
	1.24 kg H_2O	1.93 kg H_2O	3.21 kg H_2O
Secondary combustion products, including products after chemical reactions in atmosphere	$NO_x \to O_3$	$NO_x \to O_3$	$NO_x \to O_3$
	$HC \to O_3$		
	$CO \to O_3$	$CO \to O_3$	
	$SO_2 \to H_2SO_4$		
	particles		
	soot		

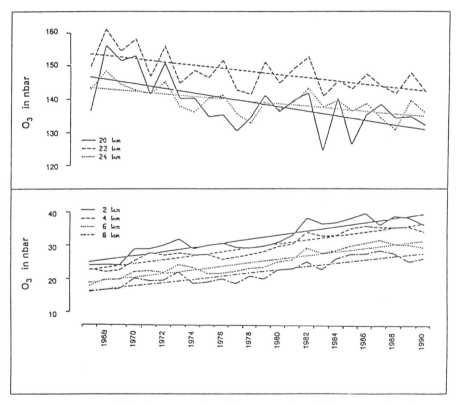

Fig. 2.6 Ozone trend for the troposphere (lower panel) and the stratosphere (upper panel) as measured at the station Hohenpeißenberg, Germany.

emission in Germany could be stabilized during the last few years the contribution by traffic has been constantly increasing (Fig. 2.7).

2.2.2 Pollutants with medium lifetime

Carbon monoxide (CO) with a lifetime of weeks to a few months pollutes continents and nearly an entire hemisphere. Since its concentration increases with roughly 1% per year in the northern hemisphere, CO has changed air chemistry on a very large scale, because it is also involved in photochemical smog reactions. Part of the observed continent-wide increase in tropospheric ozone over America and Europe during the last decades (see Fig. 2.6) is attributed to CO pollution, which has as one of its main sources traffic exhaust from engines without catalytic converters.

Also water vapour is an exhaust gas with medium lifetime in the atmosphere. If emitted at the surface, it is deposited after one to two weeks by precipitation. Its contribution to natural water vapour concentration is, however, negligibly small. If emitted from an airplane above about 8 km height it becomes an ecologically significant substance through both artificial clouds (contrails) and a potentially significant increase of natural water vapour concentration at these altitudes with normally less than a thousandth of the surface water vapour concentration.

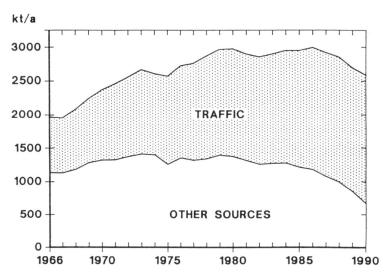

Fig. 2.7 Nitric oxide emissions in Germany, without new federal states [3].

2.2.3 Long-lived pollutants

The main exhaust of traffic, carbon dioxide (CO_2), is a very long-lived trace gas in the atmosphere. The anthropogenic CO_2 burden survives at least one century before its concentration is reduced considerably through uptake into the ocean. Therefore, the present day CO_2 emissions result in a constantly increasing atmospheric concentration of about 0.4% per year (see Fig. 2.1) which can only be reduced by rather drastic measures (see Fig. 2.4).

2.2.4 Peculiarities of air traffic

Aircraft mainly emit into the upper troposphere and lower stratosphere. Thus their environmental impact is enhanced in a manifold way:

- Most trace gas concentrations fall off with height more rapidly than pressure or density (which is roughly one quarter of the surface value at a cruising altitude of 11 km). Therefore, the relative addition near the tropopause is higher than at the surface. This is especially important for water vapour as it makes water vapour emissions by aircraft environmentally relevant.

- The residence time of an admixture in cruising altitude is longer than for near surface emission. It increases, for example, from about one week to months for water vapour.

- Very low temperatures below $-55\ °C$ allow under certain meteorological conditions rather persistent thin ice clouds (contrails).

- The radiation budget change caused by an admixture reaches a maximum at tropopause level because of the low temperatures there.

NO_x emission by aircraft contributes to ozone formation more effectively than emission into the lowest atmosphere by cars. Ozone, contrails and water vapour at low temperatures near to or at the tropopause all enhance the greenhouse effect of the atmosphere much more strongly than equal additions at higher temperatures. For ozone the radiation budget change per unit mass addition is by a factor of 1000 to 2000 higher than for carbon dioxide at 10 to 15 km height. For water vapour this factor reaches 100 in the lower stratosphere (Fig. 2.8). Due to the rapid decrease of saturation water vapour density (and, thus, also the actual vapour density) with decreasing temperature (about 1/5 per degree) anthropogenic input becomes considerable at low temperatures in the upper troposphere and lower stratosphere. But as water vapour is the most important greenhouse gas of

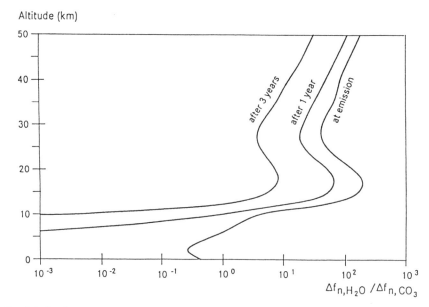

Fig. 2.8 Both water vapour and CO_2 are effective greenhouse gases, changing the global radiation balance if increased. Shown is the change in thermal radiation net flux at the top of the atmosphere (mid-latitude summer atmosphere) for water vapour related to the effect of the same number of molecules of CO_2 injected initially. As water vapour is rapidly removed from the atmosphere, the importance of an initially emitted amount reduces with the time as compared to the long-lived CO_2.

the natural atmosphere, any significant anthropogenic addition increases this effect considerably. Also the increase in greenhouse forcing by any substance is largest when it is released near the coldest level of the atmosphere, because there it will shield the radiation from the (warm) surface or lower layers most effectively. Only a 10% increase in water vapour concentration between 10 and 15 km would increase the greenhouse effect by the same amount as an 18% increase in CO_2 [5].

On the other hand, input of water vapour at small ambient vapour concentration increases the chances for the development of long-lived ice cloud areas out of initial contrails (Fig. 2.9). Their climate impact is also likely to increase the greenhouse effect. Estimates of this are not very reliable today but vary between a surface temperature increase between 0.1° and 1° per 1% increase in global cloudiness. This variability results from the hitherto barely known optical properties. Also the global cloudiness change due to contrails is not known today. Preliminary results for the European sector show a 0.5% increase in average cloudiness, with local maxima in

28 ENERGY AND THE ENVIRONMENT

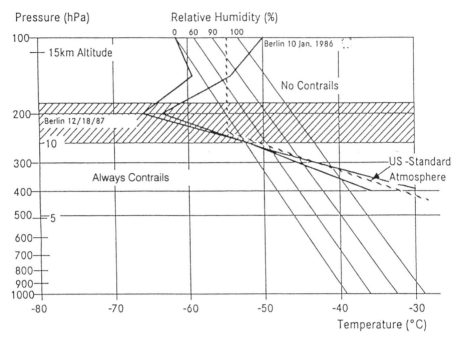

Fig. 2.9 When environmental temperatures fall below the indicated limits (which depend to some extent on humidity) contrail formation becomes probable. This is typically the case near the tropopause when temperatures are below −50°.

spring and summer over the north Atlantic route of more than 2% (Fig. 2.10).

2.2.5 Consequences for traffic fuels

Carbon dioxide emission reduction is the main consequence of the UN Climate Convention, and it will be the driver of all environmental measures concerning fossil fuels in the years to come. Increasing fuel efficiency of cars, trucks, ships and other vehicles would always mean improving the environmental quality in many respects, because NO_x, CO, hydrocarbon and toxic trace substance emissions would be lowered simultaneously with the CO_2 emissions.

Hitherto, environmental protection mainly meant reduction of certain locally or regionally important pollutants by end-of-the-pipe technologies, a prominent example being the catalytic converter for cars. Since all emissions

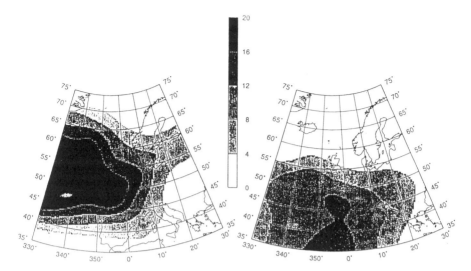

Fig. 2.10 Increase in cloud coverage by contrail generated cirrus fields in permille for summer (June–August, left panel) and winter (December–Feburary, right panel) derived from the analysis of six years of NOAA satellite images (1979–1981, 1989–1991).

from fossil fuel burning are potentially harmful, a new approach to environmentally safer technologies is necessary.

The best approach for the transportation sector from the point of view of the environment would be regenerative fuels, for instance trains powered by hydroelectric current, regenerative hydrogen fuelled, or electric, cars and buses.

2.3 ALTERNATIVE ENERGY PRODUCTION AND FUELS

The facts mentioned in section 2.2 urge the transition to environmentally less threatening energy production technologies. Various possibilities have been proposed, all of which have their own environmental and logistical problems. For the assessment of environmental acceptance the whole life cycle of production – transport – combustion has to be considered.

2.3.1 Alternative fuels

Natural gas, consisting mainly of methane (CH_4) could replace the classical gasoline and/or kerosene within a rather short time span without too much change in the overall fuel handling and distribution systems. Table 2.1 lists the type and amount of the relevant combustion products. Due to a higher energy content, the specific CO_2 emission is a few percent less than that of gasoline or kerosene while the other emissions remain roughly the same. It is known, that today pipeline transport of natural gas results in considerable leakage which may make up for about 10% of the global methane sources. This is of immense concern, as methane is about 30 times more effective in increasing the global greenhouse effect than CO_2 is, and as the atmospheric concentration of methane is presently increasing by nearly 1% per year. Also, natural gas, is a limited resource, which has to be replaced by alternatives at some time in the future. Nevertheless, it may be worthwhile to be considered for a transition phase to sustainable technologies.

Fuel from biomass is already widely used in some regions of the world. Either alcohol or artificial hydrocarbons are refined from certain plants and admixed to chemical fuels. This procedure is nearly neutral to the CO_2 budget, as energy consumption for the production and distribution of these fuels is low compared to the energy content itself. Depending on the specific fuel types various impurities result in specific environmental impacts, that have to be assessed for each case. The potentially available amount of these fuel additions is very much limited by the available land for biomass growth. High fertilizer usage is known to result in the emission of N_2O, another very effective greenhouse gas. Therefore, this technique will probably remain restricted to certain regions and niche markets.

2.3.2 Electric power from alternative sources

Another alternative in energy supply to traffic is the direct provision of electric energy from power generators either by transmission through wires and cables or by batteries and/or accumulators by:

- classical coal, oil, gas fired power plants;
- atomic power plants;
- fusion power plants;
- wind, wave, and solar power plants.

The electric power is partly absorbed in the wires before it reaches the user and is therefore lost. These wire losses contribute to the energy consumption in Germany with a few percent only. Batteries and accumulators are made of various combinations of rare elements, some of which are very dangerous for the environment. Lead accumulators or cadmium in batteries are prominent examples. Huge amounts of these materials have to be produced, handled and dumped if considerable energy amounts are to be transported. As this is too expensive in comparison with most alternatives batteries and accumulators will always be restricted to niche applications and, thus, to rather small energy portions.

2.3.3 Hydrogen as an alternative fuel

Hydrogen has been proposed as an almost ideal means of storing energy and providing it for traffic purposes. Ideally, production is possible from water with the energy available from alternative (e.g. solar, hydroelectric) power plants. Unfortunately, the regenerative production of fuels does not avoid environmental impact totally, it just reduces it. Let us assume that cryogenic hydrogen for airplanes or hydrogen for cars is derived from hydroelectric power plants which almost completely avoids CO_2 emissions. Also the emissions of CO, hydro-carbons, SO_2, and soot particles are negligible. The combustion of hydrogen, however, still causes the emission of nitrogen oxides (NO_x) depending on temperature and temperature gradients within the engine. The consequence would be a changed tropospheric chemistry and thus an increase in ozone. Therefore, measures to reduce NO_x emissions have to be considered already during the research and development phase of the future fuel in order to reduce the environmental impact nearly completely.

Hydrogen combustion in a heat engine causes water vapour as the main exhaust. In stoichiometric combustion the production of the same energy amount results in the emission of about three times more water vapour than from fossil fuel burning. This will not affect the natural water vapour concentration and distribution considerably, as long as the natural water vapour concentration is high which is the case in the warmer (lower) parts of the troposphere. But at an ambient temperature below −45 °C, that means in typical cruising heights of present day aircraft, the impact of the additional water vapour is considerable. Besides an increase in greenhouse potential (Fig. 2.8), an increase in the amount of contrails with a potentially large impact on the Earth's radiation balance has to be anticipated (see also section 2.2.4).

Therefore, a flight level below the tropopause is required in order to avoid accumulation of exhaust products in the stable lower stratosphere. If the ambient temperature is below −45 °C an additional reduction of the flight level is required to avoid contrail formation. Only 1000 m reduction in flight level below the tropopause offsets the increased water vapour emission of H_2 – as opposed to kerosene powered aircraft. Also at such a level the additional greenhouse effect of the additional water vapour is reduced below that of the additional CO_2 of an equivalent kerosene aircraft (Fig. 2.11).

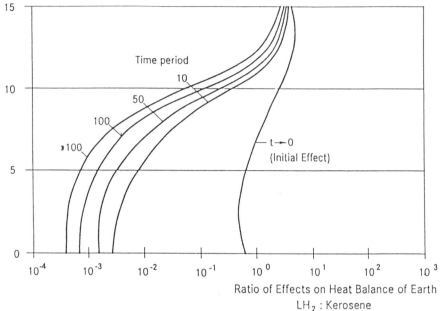

Fig. 2.11 Ratio of the contribution to the change in global radiation balance by the exhaust products of hydrogen combustion and kerosene combustion for the same energy amount. The change with time results from rapid removal of water vapour from the troposphere as compared to CO_2 [4].

2.4 CONCLUSIONS

Conventional fuels in ground and air transportation contribute an essential and intolerable amount to the environmental deterioration from mankind's energy production. The ultimate goal of a stabilization of the atmospheric CO_2 level calls for strong measures to reach about 1% emission reduction per year, to which all fields of energy production will have to contribute. This will mean the gradual shift to a different fuel for traffic that avoids the CO_2 emission.

None of the possible alternatives is completely neutral to the environment, but considering all aspects hydrogen produced from clean sources (e.g. solar or hydropower) seems to be the most promising fuel for the long-term future. Of course, measures must be taken to minimize NO_x release, in order to reduce the anthropogenic ozone production in the troposphere. Also in air traffic changes in flight levels to below the tropopause and eventually to levels with temperatures above $-45\,°C$ are necessary to reduce the climate impact to below that of conventionally fuelled aircraft. Quantitative details of these changes and the underlying environmental impacts will have to be subject of additional research.

2.5 REFERENCES

1 IPCC, *Climate Change*, The IPCC Scientific Assessment, Cambridge University Press, Cambridge, 1990, pp. 365.
2 UNCED, *UN Framework Convention on Climate*, United Nations, New York, 1992.
3 Umweltpolitik, 'Verminderung der energiebedingten CO_2-Emissionen in der Bundesrepublik Deutschland', Deutscher Bundestag, Drucksache 12/2081, 1992.
4 H.G. Klug, H. Graßl, Aircraft Using Cryogenic Fuel and Their Impact on the Atmosphere, *European Geophysical Society, XVIII General Assembly*, Wiesbaden, 1993.
5 C. Brühl, P. Crutzen, E.F. Danielsen, H. Graßl, H.-D. Hollweg, D. Kley, 'Umweltverträglichkeitsstudie für das Raumtransportsystem SÄNGER', Max-Planck Institut für Meteorologie, Hamburg, 1991 pp. 142.

3
Characteristics of Alternative Fuels – Limiting the Choice

H.G. Klug, H. Buchner
Deutsche Aerospace Airbus, Hamburg, Germany
Daimler-Benz Central Research, Stuttgart, Germany

3.1 INTRODUCTION

The question of selecting a suitable energy carrier for future transportation systems must be considered under the aspects of:

- energy source: which energy carrier can be produced from acceptable energy sources?
- vehicle: which energy carrier results in acceptable vehicle characteristics?

3.2 ENERGY SOURCES AND ENERGY CARRIERS

Today, crude oil, natural gas, and coal are the fossil basis of our energy system. Crude oil is used to produce gasoline, gas oil, and kerosene, the fuels used in ground and air transportation.

Oil shale, oil sands, bitumen can be exploited in the future. Natural gas and coal can be used to produce synthetic fuels. However, using these sources implies higher cost and higher environmental penalties as the production and conversion processes are more complicated and need more energy.

In any case, using these energy sources and the energy carriers produced from them, respectively, increases the CO_2 content of the atmosphere.

(A possible exception will be mentioned in a later chapter.) It is exactly these energy sources/carriers for which an alternative is sought.

Nuclear power can today only be produced by fission reactors (using the limited Uran resources). Fusion reactors are still in the far future. The energy carrier produced primarily is electricity, which can be converted into the chemical energy carrier hydrogen by electrolysis.

New, non-fossil, regenerative or practically unlimited energy sources like water power, tidal power, ocean wave power, geothermal power will be neutral to the atmosphere and will supplement, and hopefully replace in the long run, the fossil energy sources. All these clean energy sources primarily yield electricity, which can be converted by electrolysis, into the chemical energy carrier hydrogen.

Biomass is also a regenerative energy source. Chemical energy carriers – alcohols, gas oil-type, and maybe even kerosene-type fuels – can be produced from biomass through biochemical and chemical processing. As the carbon contained in the fuels originally was collected by the plants out of the atmosphere, biomass can be considered as a clean energy source (at least with regard to CO_2). However, raising fuel crops competes with food production for the growing human population, and hence cannot be the universal answer.

So, when considered from the point of energy source, hydrogen appears to be an ideal energy carrier, as it can be produced from literally every energy source. Furthermore, it can easily be stored and distributed, which is very important in view of the natural fluctuations in the availability of many of the regenerative energy sources.

3.3 CHARACTERISTICS OF ENERGY CARRIERS

Rail-bound vehicles can draw the required energy from electric power lines continuously. Road vehicles, aircraft, and ships must carry along the energy they need for propulsion stored by some means.

Energy can be stored:

- mechanically (e.g. fly wheels);
- electrically (batteries);
- chemically (fuels e.g. hydrogen, hydrocarbons).

The most important characteristic of an energy storing medium or device is its energy density, both related to mass and to volume. Figure 3.1 gives energy densities of alternative energy carriers.

		kJ/kg	kJ/l	Remark
Chemical Fuels - all liquid	Gasoline	$4{,}45 \times 10^4$	$3{,}25 \times 10^4$	
	Kerosene	$4{,}28 \times 10^4$	$3{,}42 \times 10^4$	
	Natural Gas	$4{,}8 \times 10^4$	$2{,}0 \times 10^4$	Liquified
	Hydrogen	$12{,}0 \times 10^4$	$0{,}852 \times 10^4$	Liquified
	Methane	$5{,}0 \times 10^4$	$2{,}1 \times 10^4$	Liquified
	Ethane	$4{,}75 \times 10^4$	$2{,}59 \times 10^4$	Liquified
	Propane	$4{,}64 \times 10^4$	$2{,}71 \times 10^4$	Liquified
	n-Butane	$4{,}58 \times 10^4$	$2{,}75 \times 10^4$	Liquified
	Methanol	$1{,}95 \times 10^4$	$1{,}54 \times 10^4$	
	Ethanol	$2{,}69 \times 10^4$	$2{,}14 \times 10^4$	
	Lead acid battery	90,0	470	
	High temperature battery	430,0	250	

Fig. 3.1 Energy density.

3.4 CRITERIA FOR SELECTION OF ENERGY CARRIER

Ideally, a new energy carrier for non-rail-bound vehicles should meet the following requirements:

- it should not rely on a limited fossil basis;
- it should be based on energy sources/raw materials which are available worldwide, not limited to few countries;
- it should not rely on any specific energy basis to allow for every possible mix of energy sources;
- its production should not compete with other basic needs of mankind, such as food;
- energy density per mass and per volume must be high not to impair the mission of the vehicle;
- operation should be easy;
- safety must be as important as for today's systems;
- CO_2 content of the atmosphere must not be increased;
- there should be no harmful emissions or waste;
- cost should be low.

38 CHARACTERISTICS OF ALTERNATIVE FUELS

Relative importance of the requirements can vary between different kinds of vehicles.

3.5 SELECTION OF ENERGY CARRIER FOR AIRCRAFT

The aircraft is an extremely weight sensitive long range vehicle. Energy content per mass is of primary importance in selecting an energy carrier. Nothing of lower energy density per mass than kerosene is acceptable. This excludes the alcohols. Batteries are completely out.

Figure 3.2 compares the alternative fuels with the criteria mentioned above. Obviously, for aircraft there are only two promising alternative fuels/ strategies:

1. Develop a method to produce kerosene (or a fuel as close to kerosene as possible) from biomass. Nothing is easier to handle than kerosene, no change in aircraft/airport would be required. Whether this is feasible is not known at the moment. In any case, it appears to be impossible to produce the full amount of fuel required this way. Furthermore, burning bio-hydrocarbons may be neutral to the atmosphere in terms of CO_2

	Kerosene, fossil basis (reference)	Kerosene, biomass basis	Liquified natural gas	Liq. hydrogen, regen. energy basis	Liq. ethane, propane, butane, biomass basis	Methanol, biomass basis	Ethanol, biomass basis
Not relying on fossil basis		✓		✓	✓	✓	✓
Universally available		✓		✓	✓	✓	✓
No specific energy basis			✓				
No competition to other basic needs	✓		✓	✓			
Energy density per mass	✓	✓	✓	✓✓	✓		
Energy density per volume	✓	✓					
Easy operation	✓	✓				✓	✓
Safety	✓	✓	✓	✓	✓	✓	✓
No increase of CO_2 in atmosphere		✓		✓	✓	✓	✓
No/little harmfull emissions/waste			✓	✓	✓	✓	✓
Low cost	✓		✓				

Fig. 3.2 Comparison of alternative fuels.

emission, but there will still be the emission of CO, unburnt hydrocarbons, and NO_x.

2. Develop the technology for producing, storing, and distributing great amounts of liquid hydrogen (LH_2), and for using it safely and economically in aircraft. The high energy density per mass promises payload/range advantages over kerosene. The main disadvantages are the low density, and the more complicated systems and operational procedures required. However, LH_2 is the only fuel which can be produced everywhere, on the basis of every clean energy source. Hydrogen is produced from water by electrolysis, and water is the combustion product, so there is a closed cycle. Still, as the water is emitted by the aircraft at altitudes of usually very dry air, the environmental effect needs careful study. NO_x still can be produced, so it must be a priority target to develop the technology to avoid its formation.

In terms of the very important parameter 'energy per mass', natural gas (which is mainly methane) looks promising. However, it is a fossil fuel, and the environmental effect of burning it is not much better than the effect of kerosene.

Nevertheless, it must be considered seriously. If a country runs out of its fossil fuel basis (crude oil), while not sufficient regenerative energy is produced yet, some in-between solution may become necessary. For a country like Russia with large reserves of natural gas, the use of liquified natural gas (LNG) as an aviation fuel may be an acceptable stepping stone to the final solution LH_2.

Extracting the hydrogen from the natural gas without releasing CO_2 into the atmosphere would be an attractive variant. Such processes are under study and will be mentioned in a later chapter.

3.6 SELECTION OF ENERGY CARRIER FOR ROAD VEHICLES

Besides gasoline and gas oil, there is a number of liquid or gaseous hydrocarbons (methanol, ethanol, biofuels, propane, butane, natural gas etc.) which can be used as a fuel for combustion engines. To some degree, they even could contribute to the reduction of harmful emissions caused by the traffic.

However, if zero emission vehicles are required (as e.g. by the Californian emission laws), only hydrogen-fuelled combustion engines (lean mixtures to avoid NO_x!) or electric motors (based upon batteries or hydrogen fuel cells) are feasible. The relative merits will be discussed later; however, a glance at

Fig. 3.1 already indicates the weight problem which goes with the use of batteries.

As the primary objective of anti-emission laws of the Californian type is the local reduction of emissions, the source of the hydrogen is of secondary importance in the beginning.

Later, when more and more regenerative energy sources will be utilized to generate electricity, hydrogen can be produced by electrolysis. Emissions thus can be reduced on a global scale, not just locally.

Of course, road vehicles can also be powered by biofuels. However, the argument given above against the use in aviation is even more true for use in road vehicles: it will not be possible to produce the required quantities. Besides, a bio-hydrocarbon powered car does not have zero emissions (CO, unburnt hydrocarbons, NO_x).

Thus, hydrogen is seen as holding the biggest long-term attraction for road vehicles, too.

4
Clean Energy Sources

R. Wurster

Ludwig-Bölkow-Systemtechnik, Ottobrunn, Germany

4.1 INTRODUCTION

'Clean energy sources' shall be defined as energy sources which can provide a sustainable and environmentally compatible energy supply. Such a type of energy supply has to take into account the environmental constraints and the safety considerations, and the appropriate energy needs of a rapidly growing world population.

In the long run, these requirements cannot be fulfilled by fossil fuels (aggravating the greenhouse effect, local pollution) or by nuclear energy (unsolved fuel waste problem, accident potential, proliferation problem), since their use would have to be considerably extended.

Therefore, renewable energy sources seem to provide the only long-term option which are sustainable. They will also acceptable by the future generations which do not have a vote today. Already today, some 16,000 TWh of thermal energy (1400 mill.toe) are produced from non-commercial biomass worldwide (see Fig. 4.1), covering about one seventh of world primary energy consumption (of approximately 400 EJ) or representing almost the fourfold coverage as provided by nuclear power in 1985 [49], [54]. In 1985, worldwide some 10,000 TWh of electricity were generated and some 220 EJ of energy were consumed in direct fuel use [54].

As renewable energies are regarded as concepts using the hydrological cycle (hydropower), the gravity (wave power, tidal power), differences in water temperature (ocean thermal energy conversion), movement of air (wind energy), direct solar use (solar thermal energy conversion, photo effect), and indirect use of solar energy via biomass. Geothermal energy is not regarded as a really renewable form of energy and therefore is not considered in this overview, although its potential may be significant.

CLEAN ENERGY SOURCES

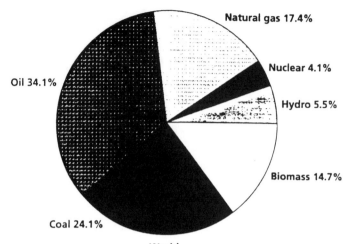

World
Total = 373 exajoules
Population = 4.87 billion
Energy use per capita = 77 gigajoules

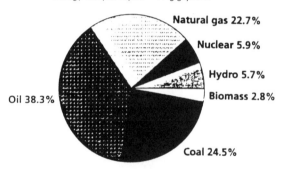

Industrialized countries
Total = 247 exajoules; 66% of world total
Population = 1.22 billion; 25% of world total
Energy use per capita = 202 gigajoules

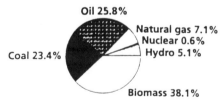

Developing countries
Total = 126 exajoules; 34% of world total
Population = 3.65 billion; 75% of world total
Energy use per capita = 35 gigajoules

Fig. 4.1 Primary energy use for the world (top), industrialized countries (middle) and developing countries (bottom) in 1985 [54].

4.2 REVIEW OF RENEWABLE ENERGY POTENTIALS

4.2.1 Hydropower

The present hydropower utilization worldwide amounts to a total of about 2000 TWh/year of electric energy (1988). The technically exploitable hydropower potential lies around 17,000 TWh/year (estimates for the technically exploitable hydropower potential range between roughly 15,000 TWh/year [13] and 19,000 TWh/year [2]), which is about eight times the present electricity generation from hydropower or almost one quarter of the world primary energy consumption of 1987. Roughly a quarter of the potential is located in each, Africa, Asia and Latin America respectively. The remaining quarter is attributable to Europe, North America and Oceania. These estimates do not include the technically exploitable potential of Greenland which is estimated to be in the order of 1500–1800 TWh/year [2, 11]. Due to possible environmental impact produced by large-scale hydropower utilization these potential estimates probably have to be reduced by about 35% [44] in the long run, thus coming to then exploitable potentials of slightly above 10,000 TWh/ year. Other sources [54] derive the utilizable potential from the rate applicable to industrialized countries and come to 6000–9000 TWh/year. Other reasons why existing potentials might not be opened are scarcity of funding and the lack of nearby markets for the electricity [49]. From the present hydroelectricity production of 2000 TWh/year [13] (which in OECD countries equalled electricity production from nuclear energy in 1985 [49]) an increase of more than 50% to slightly over 3000 TWh/year [2] will take place until the year 2000. For the year 2020 even 8000 TWh/yr are forecast by the World Energy Conference [49].

The remaining global potential for energy trade (excluding possible scenarios in which 100% of the electricity demand is covered) is considered to be in the range between 6500 and 10,500 TWh/yr (in 2010), which is equivalent to a capacity in the range between 1400 and 2300 GW. The remaining potential calculated for the year 2040 is smaller than that for 2010 because electricity demand will increase steadily. It has been estimated to be about 4800 to 9500 TWh/yr (1070 to 2100 GW) [16].

4.2.2 OTEC

Ocean thermal energy conversion (OTEC) [15, 16] uses the solar energy stored in the surface waters of the tropical oceans and is based on the principle that energy can be extracted by linking two water reservoirs of different temperatures. The radiation warms the water in many regions of the oceans within in depth of 50 to 100 meters to a temperature of 27° to 30 °C, but at a depth of 1000 m the temperature ranges from 3° to 6 °C. Thus a typical temperature difference of 21 to 27 K is available 24 hours per day, and 365 days a year for OTEC power generation. Already a temperature difference as small as 20 K can be exploited effectively to extract usable energy.

The operating principle of closed-cycle OTEC is a working fluid cycle powering a turbine. The turbine drives a generator, in order to produce electricity. The opencycle OTEC system works with warm seawater as working fluid. The warm seawater is flash evaporated under vacuum in order to generate steam at about 2.4 kPa pressure. The produced steam expands through a low-pressure turbine coupled to an electricity generator. The steam is condensed by cold seawater pumped from the ocean's depths through a pipe. By using a surface condenser the steam remains separated from the cold seawater and can serve as desalinated water.

The potential resources available for OTEC exploitation are immense, i.e. $> 10^{13}$ W [15] respectively $> 10^{10}$ W, and its technical potential is considered to be approximately 200,000 TWh/yr. However, in order to exploit energy on the basis of OTEC, it is necessary to shift surface waters to a depth of approximately 1000 m and this process may have severe impacts on the climate. Thus, in order to avoid climatic changes as a result of OTEC, the theoretical potential of 200,000 TWh/yr has to be limited to approximately 9000 TWh/yr, still equivalent to more than four times the present conventional hydroelectricity production worldwide [16].

OTEC represents an option for energy production that provides electricity at a steady level, thus being particularly suitable for baseload electricity generation. Furthermore, at suitable sites OTEC is considered to be an ideal method of generating electricity in order to supply isolated shorelines and remote islands which are not interconnected with a large-scale electricity grid via underwater cable. At sites where such direct electricity transmission is not possible or needed, the electricity may be used to operate a plant which produces chemicals and fuels, for instance hydrogen by electrolysis. Therefore, OTEC represents an ideal technology for the production of hydrogen, whereas hydrogen represents an ideal facilitating technology for tapping offshore OTEC potentials.

Although major efforts regarding the development of OTEC technologies have been made or are still under way in the U.S. and Japan, commercially used OTEC plants do not exist yet. With respect to market penetration, it is considered most likely that land-based and near-shore OTEC facilities with a capacity ranging in the lower MW-scale (< 15 MW) will start to operate in the near future. Near-shore and floating offshore plants are projected as being economically feasible on a medium-term basis.

Estimates on a global scale show that with intensified R&TD but without major technical breakthroughs at least 350 MW could be provided on the basis of OTEC by the year 2005, an amount which could be increased to as much as 2100 MW by the year 2010. In the longer term, the production of hydrogen (and/or methanol) for transportation fuel uses may prove to be economically feasible and could raise this potential by a factor of five to the 10 GW range. Especially 1 MW to 100 MW OTEC plants may serve a potential market in the area of the Pacific and Asian islands.

At present, based on existing designs, capital costs are in the order of US$ 10,000 per kW installed. Plants designed for unattended operation might well be able to produce electricity for 0.12 to 0.25 US$/kWh [54].

4.2.3 Wave power

The theoretical wave energy potential around the shores of the oceans is estimated in average to more than 2 TW (see Fig. 4.2). This type of energy depends very much on the wind patterns and therefore is a highly fluctuating energy source, in time as well as in magnitude. Although stronger winds usually occur in winter time and therefore coincide very often with the higher energy demands in winter, to equalize the differences in demand and supply a storage system has to be incorporated [15].

'The exploitation of the North Sea as an energy source for the production of hydrogen seems to be both technically and commercially feasible. The exigent need for renewable energy sources makes it interesting to deal with wave power' [45].

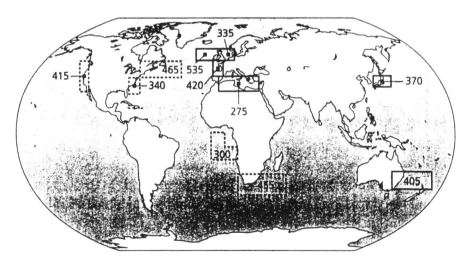

Fig. 4.2 Annual wave energy in MWh per meter from specific areas [54].

46 CLEAN ENERGY SOURCES

Various technical concepts for wave power utilization exist, such as: buoyant structures, also called heaving float devices; hinged structures, also called pitching devices which follow the contours of the waves; flexible-bag devices also called heaving and pitching float devices; structures with an enclosed oscillating water column which pump air; focusing or surge devices.

Due to the large variety of designs many cost estimates on the basis of laboratory test and some realized pilot plants in China, Denmark, India, Japan, Norway, Sweden and the United Kingdom exist. As a rough estimate for the North Atlantic where wave power is about 50 kW/m comes to 0.09 US$/kWh at an assumed discount rate of 12.5% [54].

4.2.4 Tidal energy

Tidal energy is a very predictable source of energy, varying by a factor of four between the tidal cycles. Utilization of tidal energy is practical only at selected sites with large tides (> 3 m tidal range).

In Europe the exploitable potential is estimated to be in the order of 105 TWh/year. The largest contribution could come from United Kingdom (50 TWh) and France (44 TWh) posessing about 90% of the European tidal energy potential. The world potential is estimated to be five to ten times that of Europe (500–1000 TWh/year). It is very likely that only a fraction of the possible sites (see Fig. 4.3) might be utilizable at an economic scale.

Few tidal energy plants are in operation worldwide. The plant at La Rance in France is the largest (240 MW_e and 540 GWh/year) and the oldest

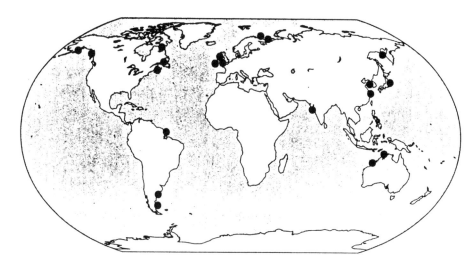

Fig. 4.3 Principal sites for tidal power development [54].

(25 years). Two other plants in the Megawatt size exist, one at Annapolis, Canada, with 17.8 MW$_e$ and 30 GWh/year output and one at Jiangxia, China, with 3.2 MW$_e$ and 11 GWh/year output. Very large plants are still in the concept stage, as e.g. the Severn barrage project which has a projected power output of 8.6 GW$_e$.

The disadvantage of tidal power plants are their long construction periods and their fluctuating operation mode, one increasing the financing costs, the other increasing the cost per kW of installed capacity due to the low load factors of between 0.22 and 0.35.

4.2.5 Wind energy

In order to make technical use of wind energy [15, 16], an average wind speed of approximately 5 m/s is necessary. The relationship between wind speed and the power obtained from it is a cubic one. This relationship is best illustrated by comparing good, excellent (good + 86%) and outstanding (good + 212%) wind sites with an average wind speed of 5.8, 7.2 and 8.5 m/s respectively. Areas with an average wind speed high enough for the technical use of wind energy can be found in Northern Canada, North-Western Europe, Eastern China, the Antarctic and coastal areas of North-West Africa and South America.

In addition to this, the offshore potentials of wind energy can also be made use of. These can be exploited in ocean waters not exceeding a depth of 5 to 6 m (a demonstration plant exists in Denmark), but the costs for operating these plants are one third higher than those for land-based wind turbines.

Wind resources vary in speed, direction and availability (see Fig. 4.4). Determining truly exploitable resources is a highly complicated and complex process and therefore parallel resource assessments for identical areas may lead to a wide range of results. In a study carried out by IIASA, the global technical potential of wind energy (excluding offshore applications) has been estimated to be approximately 3 TW or 8000 to 12,000 TWh/a. These figures show that the technical potential of wind energy could almost cover the current global consumption of electricity (1987).

Most of the regional resource assessments regarding wind energy have been carried out in Canada, the US (see Fig. 4.5) and in the European countries (see Fig. 4.6). Conservative estimates show that, in the US, accessible resources of wind energy bear the potential of providing more than ten times the electric power currently consumed in the US. A study for Europe shows that the European wind energy potential can cover three times the electric power consumption of Europe [17].

More recent investigations estimate the theoretical wind energy potential

48 CLEAN ENERGY SOURCES

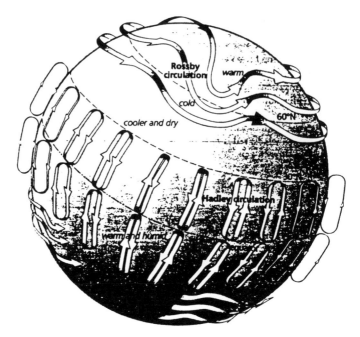

Fig. 4.4 Global wind system [54].

for electricity production at about 500,000 TWh/year (excluding offshore applications, most islands, Antarctica and Greenland). If all environmental and other yet known restrictions are imposed, a remaining exploitable potential of 53,000 TWh/year is left (see Fig. 4.7) [54].

Worldwide, some 17,800 commercial wind turbines with a capacity of approximately 1700 MW have been installed up to now, 1500 MW thereof located in California, and more than 100 MW in Denmark. In Europe and in the US, most investments in wind energy production have been subsidized by the respective governments in order to surmount the break-even point compared to competing electricity production. Since 1985, 622 MW of wind converter capacity have been installed in California without the benefit of tax incentives, in total representing a capital investment of almost 1 billion US$.

Wind energy has not only proven to be – besides hydropower – the most cost-competitive (indirect) solar energy technology commercially available, but it can also be made available for the bulk power market. The production of electricity by wind turbines is at present concentrated on industrialized countries. There exists, however, an increasing interest in developing countries in using wind energy for both the operation of water pumps and the production of electricity [18].

Fig. 4.5 U.S. wind energy resources [54].

The development of large-sized turbines is more cost-intensive and difficult than the development of smaller turbines [19]. Nevertheless, major efforts regarding the technical development and construction of large-sized turbines have been made in several countries. This is reflected in the fact that the European countries, Japan and the US make major subsidies available in order to promote wind energy RD&D. An agreement for

50 CLEAN ENERGY SOURCES

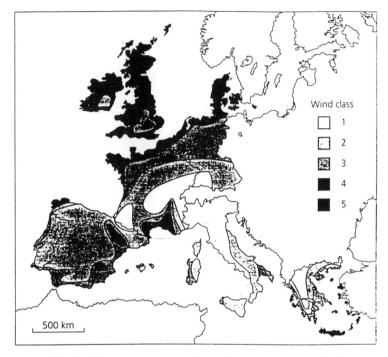

Fig. 4.6 European community wind energy resources [54].

Region	Resources class 3 and above		Population density	Estimated second-order potential
	Percent land area	Gross electric potential, TWh/year	People per km^2	TWh/year
Africa	24	106,000	20	10,600
Australia	17	30,000	2	3,000
North America	35	139,000	15	14,000
Latin America	18	54,000	15	5,400
Western Europe	42	31,400	102	4,800
Eastern Europe and former USSR	29	106,000	13	10,600
Rest of Asia	9	32,000	100	4,900
World[b]	23	498,000		53,000

(a) Numbers may not add up due to rounding up. For assumptions, see text.
(b) Excluding Greenland, Antarctica, most islands, and offshore resources.

Fig. 4.7 World wind electricity potentials [54].

cooperation regarding the development of large-scale wind conversion systems has been worked out under the auspices of the International Energy Agency.

In areas with an average annual wind speed of 8 to 9 m/s, large-sized turbines lead to generation costs of 7 to 10 ¢US/kWh (rates for consolidated technologies are 4 to 7 ¢US/kWh) and a cost reduction to 5 ¢US/kWh is presently considered achievable. In the long-term (30 years) wind could see costs reduced to as low as 3 ¢US/kWh [16].

In order to produce hydrogen on the basis of wind energy, large-sized turbines are required which themselves have to be grouped and interconnected in high numbers. These windfarms may be situated in remote areas. However, the acquisition of suitable sites may be restricted because the wind power produced in industrialized countries will generally be fed into the electricity grid. Potential sites of hydrogen production on the basis of wind power are located in West Africa, Greenland, Iceland and the southern part of Latin America (if transcontinental electricity transport tends to be too costly or inefficient).

4.2.6 Solar energy

Worldwide about 1.9 million km^2 of vast stony deserts with a global irradiance of 2000 kWh/m^{2}yr exist, representing about 1.3% of the global surface area [6, 7, 15, 20]. Depending on the level of technology applied (solar thermal power plants or photovoltaic power plants and their respective efficiencies solar radiation to electricity/hydrogen transportation via pipeline in gaseous form or containerized in liquid form/transportation distance/end-use distribution/availability of entire system) and on the assumed land use factor, between 105,833 and 138,397 TWh/yr of solar hydrogen could be generated on it. To cover the present world end-energy consumption of roughly 70,000 TWh/yr with solar hydrogen, a suitable area in North Africa of only some 850,000 km^2 (representing approximately 10% of the Sahara or little more than one third of the acreage of Algeria) would be required. In many areas of the world unused deserts and wastelands exist additionally (in Spain there are some 20,000 km^2).

Assuming a photovoltaic module efficiency of only 10% the following theoretical electricity production potentials are obtained [21]:

- Northern Europe: 88 × 10^{12} kWh/year;
- Southern Europe: 195 × 10^{12} kWh/year;
- Germany: 25 × 10^{12} kWh/year;
- Algeria: 476 × 10^{12} kWh/year.

52 CLEAN ENERGY SOURCES

In all these cases the present electricity consumption can be covered at least 50 times by the theoretically existing potential (see Fig. 4.8).

Vast areas of land in Australia, in Brazil, in North Africa, in the US and in the southern states of the former USSR can be used for large-scale solar energy conversion.

All these considerations show that land availability does not represent any problem to extensive solar electricity and hydrogen production in many areas all over the world.

POTENTIALS	ELECTRICITY GENERATION [TWh_e/year]	HYDROGEN PRODUCTION FOR FOREIGN TRADE [TWh_{th}/year]
HYDROPOWER (Remaining Exploitable Potential) (Ecologically Acceptable Exploitation)	< 22,000 ≈ 17,000 ≈ 10,000	} > 1,000 to 4,000
OCEAN THERMAL ENERGY CONVERSION (Remaining Potential)	200,000 ≈ 8,000	} > 6,000 to 7,000
TIDAL POWER	500 - 1,000	160 - 330
WIND POWER (Remaining Potential)	500,000 ≈ 50,000	} 1,000 - 5,000
DIRECT SOLAR (Remaining Potential)	> 1,000,000 ≈ 100,000	} > 20,000
BIOMASS (Remaining Potential)	83,000 ≈ 40,000	} up to 20,000[a]
TECHNICAL RENEWABLE WORLD POTENTIAL	≈ 1,800,000	
Remaining Potential for Renewable Electricity	≈ 209,000[b]	
Remaining Potential for Hydrogen Trade		> 38,000 to 56,000[c]

LBST Ludwig-Bölkow-Systemtechnik GmbH /Wu 1993-05-11

Sources: [7] [12] [15] [16] [20] [42] [54]

  a. For biomass it is assumed that half of the potential goes into electricity production and half into biomass gasification for hydrogen fuel production, of which again half is assumed to enter into international hydrogen trade.
  b. Approximately 19 to 20 times world electricity production in the late 1980s.
  c. Approximately two-thirds of worldwide 1985 direct fuel use.

Fig. 4.8 Electricity generation and hydrogen production potential for trade based on renewable energy sources.

The energy production costs for the two most promising technologies – solar thermal parabolic trough electricity generating systems (SEGS) and photovoltaic electricity generating systems (PV) – amount to about 0.08–0.12 US$/kWh for solar thermal in California (1990), whereas for utility scale PV the present costs under Californian insolation conditions have to be assumed realistically to be still in the order of 0.50 US$/kWh (1990) with the potential to drop to 0.20 US$/kWh (2000) [15].

4.2.7 Bioenergy

Biomass can be defined as wood and wood wastes, wastes from wood processing industries, charcoal, food industry waste products, sewage or municipal solid waste products, manure from animal breeding, biological materials cultivated as energy crops, industrial bio-wastes, landfill gases, etc.

The technology for ecological, energetic conversion and utilization is basically available and continuous improvements are under way.

Biomass, in its various forms, is used preferably either in direct combustion processes for the production of thermal energy for direct use or for electricity production, or it is converted bio-chemically into gaseous (e.g. anaerobic gasification) or liquid form (ethanol, methanol, synthetic hydrocarbon fuels).

Gasification of wood, peat, dry residues and miscanthus senensis to lean gas or syn-gas and its later shift reaction to hydrogen has either been proven technically or is in pilot state.

Another possible conversion path is photobiological hydrogen production which has not yet emerged from laboratory state.

For hydrogen production from biomass, the raw product preferably should be available in liquid or gaseous form for synthesis.

In 1987, worldwide some 55 EJ of energy were obtained from biomass, representing 14% of a world energy consumption of almost 400 EJ. The percentage was more than 10 times higher in developing countries (35%) as in industrial countries (3%) [46].

The technical potential for biomass in Germany (West) is estimated to be in the order of 0.35–0.43 EJ, out of which some 0.13–0.18 EJ might be used until 2005, respectively some 0.28–0.36 EJ until 2050 [47].

The technical world biomass potential e.g. on the basis of plantation of energy crops can be estimated at approximately 8 billion toe [335 EJ] under the assumption of yields of 2 toe [83.8 GJ]/ha·yr from 40 million km^2 of plantation area. Woody solid fuels, vegetable oils and ethanol are the best known and developed forms of bioenergy. Results from various climatic world regions are available and have proven energy yields of in the order of

Region	Commercial energy use[c]	Recoverable residues				Biomass plantations[d]
		Crop[e]	Forest[f]	Dung[g]	Total	
Industrialized						
US/Canada	87.9	1.7	3.8	0.4	5.9	34.8
Europe	79.8	1.3	2.0	0.5	3.8	11.4
Japan	16.6	0.1	0.2	-	0.3	0.9
Australia + NZ	3.6	0.3	0.2	0.2	0.6	17.9
Former USSR	56.9	0.9	2.0	0.4	3.3	46.5
Subtotals	*244.8*	*4.3*	*8.1*	*1.6*	*14.0*	*111.5*
Developing						
Latin America	17.4	2.4	1.2	0.9	4.5	51.4
Africa	9.2	0.7	1.2	0.7	2.6	52.9
China	23.0	1.9	0.9	0.6	3.4	16.3
Other Asia	27.7	3.2	2.2	1.4	6.8	33.4
Oceania	-	-	-	-	-	1.4
Subtotals	*77.3*	*8.2*	*5.5*	*3.6*	*17.2*	*155.4*
World	**322.1**	**12.5**	**13.6**	**5.2**	**31.2**	**266.9**

a. All energy values are expressed on a higher heating value basis.

b. Regional residue estimates are aggregates of country-by-country inventories compiled at the Information and Skills Centre of the Bioenergy Users' Network, King's College, London [1].

c. Commercial energy use for 1985, as estimated by the US Department of Energy [86], excluding biomass.

d. It is assumed that plantations having an average yield of 15 dry tonnes (with a heating value of 20 gigajoules per tonne) per hectare per year are established on 10% of the total amount of land now in forests/woodlands + cropland + permanent pasture—some 372 million hectares in industrialized countries and 518 million hectares in developing countries (see table 13).

e. Included are residues from cereals, vegetables and melons, roots and tubers, sugar beets, and sugar cane. Not included are residues from pulses, fruits and berries, oil crops, tree nuts, coffee, cocoa and tea, tabacco, or fibre crops. Crop production data are from FAO sources [66, 83]. Except for sugar cane it is assumed that 1/4 of all residues generated are recoverable. For sugar cane it is assumed that all bagasse is recoverable and that 1/4 of the tops and leaves are recoverable (see note c, table 2). For crops other than sugar cane the following are the assumed residue generation rates per tonne of crop, and the assumed heating values of these residues:

	Residue generation rate	Residue heat content
	tonnes per tonne	gigajoules per tonne, air-dry basis
Cereals	1.3	13
Vegetables and melons	1.0	6
Roots and tubers	0.4	6
Sugar beets	0.3	6

f. It is assumed that three-fourths of the milling and manufacturing wood wastes and one-fourth of the forest residues are recoverable (see notes e and f, table 2).

g. It is assumed that 1/8 of the dung generated is recoverable (see note d, table 2).

Fig. 4.9 Commercial energy use and potential supplies of biomass for energy (EJ per year) [54].

3000 litres oe/ha·yr to 10,000 litres oe/ha·yr for annual and perenial crops [51].

Under other assumptions, assuming a yield of 15 dry tons per hectare and year (upper heating value of 20 GJ/ton and a utilized land area – 10% of the total area in agricultural/forestal use – of 372 million ha in industrialized and 518 million ha in developing countries, thus in total some 8.9 million km^2) biomass plantations have a potential of 267 EJ/year (see Fig. 4.9). Recoverable residues from crops, forests and dung can contribute another 31 EJ/year, thus increasing the total biomass potential for energy supply to slightly above 300 EJ/year [54].

Present electricity costs from biomass-fired steam-electric plants in the 30 MW$_e$ size on the basis of a2 biomass price of 3 US\$/GJ are in the range of 0.08–0.10 US\$/kWh$_e$ [54].

4.3 MAJOR DEVELOPMENT TRENDS AND R&TD UNDERWAY

4.3.1 Hydropower

Hydropower is a proven technology which is in large-scale commercial use for more than a century (see Fig. 4.10). In general no major improvements are expected for turbine technology and efficiency and its cost. Nevertheless new turbine designs for low-head application are being developed.

One of the major improvements in construction technology is roller compacted concrete, which allows for faster construction and thus reduces construction costs at increased durability, strength and permeability. A step-by-step replacement of gravity dams by this technology seems foreseeable.

Due to the remote location of most large-scale hydropower schemes the development of efficient electricity transmission technologies was spured, especially of high-voltage-direct-current-transmission technology (HVDC) and of very high voltage (> 765 kV) alternating transmission lines and transformer equipment.

4.3.2 OTEC

The technology is proven only at a small scale of some 100 kW. Its larger application suffers from poor economics mainly attributed to the low efficiency of the heat-power cycle, requiring large-dimensioned components handling the huge water throughputs necessary. Nevertheless it seems that

LARGEST GENERATING CAPACITY	Year of initial operation	Country	Rated capacity now megawatts	Rated capacity planned megawatts
1 Itaipu	1983	Brazil–Paraguay	10,500	12,600
2 Guri	1986	Venezuela	10,300	10,300
3 Grand Coulee	1942	United States	9,780	10,830
4 Krasnoyarsk	1968	Russia	6,000	6,000
5 La Grande 2	1979	Canada	5,328	5,328

HIGHEST EMBANKMENT DAMS	Year completed	Country	Type[b]	Height metres
1 Nurek	1980	Tajikistan	E	300
2 Chicoasén	1980	Mexico	E/R	261
3 Guavio	1989	Colombia	E/R	243
4 Mica	1972	Canada	E/R	242
5 Chivor	1975	Colombia	E/R	237

HIGHEST CONCRETE DAMS	Year completed	Country	Type[b]	Height metres
1 Grande Dixence	1961	Switzerland	G	285
2 Inguri	1980	Georgia	A	272
3 Vajont	1961	Italy	A	262
4 Mauvoisin	1957	Switzerland	A	237
5 El Cajón	1985	Honduras	A	234

LARGEST VOLUME DAMS Construction material	Year completed	Country	Type[b]	Volume 10^6 m^3
1 Tarbela	1976	Pakistan	E/R	148.5
2 Fort Peck	1937	United States	E	96.0
3 Tucurui	1984	Brazil	E/G/R	85.2
4 Guri	1986	Venezuela	E/G/R	78.0
5 Oahe	1958	United States	E	70.3

LARGEST CAPACITY RESERVOIRS	Year completed	Country	Type[b]	Reservoir volume 10^9 m^3
1 Owen Falls[c]	1954	Uganda	G	2,700.0
2 Bratsk	1964	Russia	E/G	169.0
3 Aswan High	1970	Egypt	E/R	162.0
4 Kariba	1959	Zimbabwe–Zambia	A	160.3
5 Akosombo (Volta)	1965	Ghana	R	148.0

a. [16, 17].
b. Dam type: A = arch; E = earthfill; G = gravity; R = rockfill.
c. Major part of lake volume (Lake Victoria) is natural.

Fig. 4.10 The world's largest dams and hydro plants in operation in 1989 [54].

potential improvements in the heat exchangers and in the coldwater pipe are not improving overall economics significantly. Applications contributing a synergetic value to the mere electricity production, such as freshwater production and aquacultures, may well improve overall economics [54].

4.3.3 Wave power

Besides some small scale experiments in laboratory water tanks and at sea, also some prototype plants are in operation feeding electricity into the grid. The best medium term perspectives seem to offer shoreline or water caisson mounted devices. Near-shore and offshore concepts still suffer from insufficient knowledge on the achieveable lifetime, operating and maintenance requirements/costs. If near-shore and offshore concepts could be demonstrated successfully, the technology could enter into prototype applications within one decade from now [54].

4.3.4 Tidal energy

Major efforts presently focus on reducing capital costs and/or construction time, increasing electricity output or its value (load matching) and reducing/excluding negative environmental effects. Major improvements are expected from caissons for dam construction as well as from the utilization of diaphragm-wall construction for the whole dam or at least for ship-locks.

The double-regulated bulb turbine (as used at La Rance) is still the most favoured turbine concept, especially when pumping is required. Also pit and tube type turbines with geared generators have been developed for smaller capacities (< 25 MW$_e$). Straight-flow, rim-generator turbines as used in Annapolis are suitable for ebb-generation mode, but not for pumping mode [54].

4.3.5 Wind energy

European development programmes presently focus on wind energy converters larger than 750 kW$_e$. This development approach is important especially for Europe, where locations for wind energy production are limited (see Fig. 4.11) and significant extension of the potential might be achievable only by offshore siting.

US development approaches focus more on an incremental improvement of medium-sized wind power turbines. Major areas of activity are innovative

Country	Population density inhabitants per km²	Reduction factors from exclusions	
		First order exclusions	Second order exclusions
Contiguous US	3.14	1.6	4
Denmark	20	17	65
Netherlands	360	30	150

a. Estimates are derived from siting studies and experience in the United States, Denmark, and the Netherlands (see text). These are the only countries where land-siting surveys have been carried out in such detail to give meaningful results. First order exclusions are those that can be identified as infeasible or clearly unacceptable. Second order exclusions reflect more stringent exclusions, such as visual objections to siting.

Fig. 4.11 Estimated reduction factors due to environmental and land-use constraints [54].

airfoil development and testing, structural improvements and fatigue modelling [54].

It has not yet been decided which of the horizontal axis converter designs – the variable speed or the stall controlled one – will be the superior one in the future. For both designs energy yields will be improved by installing the blades on taller towers.

As the most important fields of R&TD activities might be named:

- advanced airfoils specially designed for wind turbines (higher energy yields, reduced loading, low noise emission);
- introduction of advanced materials and fabrication technologies;
- intelligent control concepts which are self-adapting;
- improved power electronic components;
- development of optimized siting strategies;
- design of low temperature heavy duty, low maintenance turbines (e.g. vertical axis with travelling field generator [57]).

4.3.6 Solar energy

Photovoltaics: In the case of PV the total system costs have to be reduced systematically in order to become economically successful. The cheaper the module costs become the more important will be the cost reduction of balance of system costs (BOS) (support structures, inverters, cabling). The higher the efficiency of a PV module becomes the more cost effective the entire system can be operated due to proportionally reduced BOS. Therefore, the Japanese pursue the strategy of roof top integrated systems as a first step into larger scale application of PV technology in order to reduce costs by mass production. Already conventional polycrystalline silicon PV technology can be produced at significantly lower costs than today in case optimized large-scale production lines are operated and BOS costs reduced [22, 23]. In this case even under German insolation conditions generating costs of between 0.50 and 0.70 DM/kWh$_e$ are achieveable, which under north African insolation conditions drop to about 0.30 DM/kWh$_e$.

PV thin film technologies, such as amorphous silicon or copper indium diselenide, up to now could not realize their hypothetical potential for cost reduction. This was mainly due to their still too low efficiencies or too high fabrication costs (high capital investments into production equipment) in the light of steady improvements (efficiency, production costs) in the conventional polycrystalline silicon technologies. The silicon raw material costs and the cost contribution for the fabrication of polycrystalline cells from silicon are not a very significant contributor to electricity production costs. None of the PV cell materials in use or in discussion is comparably investigated and well known as silicon (by orders of magnitude!). Therefore, a long-term R&D goal could be to execute more fundamental research into efficient, cheap and stable materials for PV cell fabrication.

A new technological concept, the nanocrystalline oxide solar cell, as developed by the Swiss Federal Institute of Technology in Lausanne claims cost reductions of more than 90% leading to module costs of 0.6 US$/W$_{e,p}$ and grid connected system costs of approximately 2.0 US$/W$_{e,p}$. The cell efficiencies at diffuse light conditions lie between 12% and 15% and shall be almost independent from temperature. The crucial question which is not yet solved is that of long-term stability. The major R&D goal for this technology is to establish long-term stability without increasing production costs.

Solar thermal electricity production: The presently most economic solar thermal energy conversion technology is the parabolic trough collector technology developed by LUZ and FLAGSOL. In order to improve the present efficiencies and thus reduce production costs, the switch from thermo-oil as an energy carrier in the superinsulated vacuum tubes has to be carried out. This transfers its heat energy by means of a heat exchanger to a secondary circuit operated with pressurized water to the so-called direct evaporation concept with just one circuit containing high pressure water vapour. This avoids the need for a heat exchanger. This step needs additonal R&D. If the perspectives for the application of solar thermal power plants

60 CLEAN ENERGY SOURCES

State	Number of facilities		Installed capacity MW_e		
	Stand-alone	Cogeneration	Stand-alone	Cogeneration	Total
Alabama	0	15	0	375	375
Arizona	2	0	45	0	45
Arkansas	1	4	2.4	10	12
California	64	30	736	255	991
Connecticut	4	3	155	14	169
Delaware	1	0	13	0	13
Florida	12	15	314	474	788
Georgia	0	5	0	36	36
Hawaii	2	13	70	129	199
Idaho	1	6	0.2	116	116
Illinois	0	1	0	2	2
Indiana	0	7	0	36	36
Iowa	2	1	11	2.2	13
Kentucky	1	1	1	1	2
Louisiana	1	12	11	300	311
Maine	4	22	88	704	792
Maryland	2	2	214	94	308
Massachusetts	2	9	38	252	290
Michigan	3	13	78	247	325
Minnesota	3	23	63	161	224
Mississippi	0	10	0	230	230
Missouri	0	2	0	60	60
Montana	2	17	18	340	358
New Hampshire	3	5	15	65	80
New Jersey	2	0	14	0	14
New York	11	17	154	425	579
North Carolina	3	27	60	351	411
Ohio	1	6	17	90	107
Oklahoma	2	1	8	17	25
Oregon	3	24	69	185	254
Pennsylvania	0	9	0	144	144
South Carolina	1	13	49	46	95
Tennessee	2	12	6	43	49
Texas	1	9	2	146	148
Utah	0	1	0	20	20
Vermont	5	3	80	218	298
Virginia	0	9	0	136	136
Washington	3	11	72	120	192
Wisconsin	5	9	55	117	172
Total	149	367	2,459	5,962	8,421

a. Based on [59]. The total here is an underestimate because the cited reference is incomplete.

Fig. 4.12 Electricity generating plants burning biomass fuels in the United States as of 1989.

improve, the manufacturer of this technology would invest into such R&D right now.

These considerations indicate that R&D can contribute to cost reduction, but large scale applications of these technologies could even accelerate the process probably more.

4.3.7 Bioenergy

By far the largest possible source of biomass for energy production is biomass from plantations. Deforested or degraded areas may benefit significantly from such plantations, ecologically and economically, and provide a secure energy source. Research and development efforts are essential in order to ensure sustainability of high biomass yields under many different geographical, social and environmental conditions. Biomass conversion (incineration, gasification, digestion, chemical conversion) is a proven technology (see Fig. 4.12) which is developed in an evolutionary way for various types of application or uses.

A first step into industrial bioenergy generation can be effected by the utilization of biological residues. After the build-up of a bioenergy industry based on bio-residues has been realized, and as soon as sustainable energy plantation concepts have emerged from R&D activities, growing demand for biomass in energy use can then be covered by biomass from plantations [54].

4.4 TRENDS IN EFFICIENCY/ECONOMICS

4.4.1 Hydropower

Hydropower plants have efficiencies (hydraulic potential converted into electricity) of between 70% and 90% which at good maintenance can be maintained over a long period (e.g. up to or more than 50 years).

Production costs of hydro-electricity mainly depend on the investment costs, which can differ considerably between 5000 ECU/kW for small and medium hydro plants in Europe and some 750–2000 ECU/kW for large-scale hydropower in Brazil, Canada or Africa. Such plants can produce electricity for costs between 0.01 and 0.04 ECU/kWh [50] (see Fig. 4.13). If only the operating costs are taken into account, these existing intermediate or large hydro plants produce electricity for only 0.002 to 0.004 US$/kWh or 10% of the aforementioned costs [54].

PRODUCTION COSTS	PRESENT [ECU/kWh$_e$]	FUTURE [ECU/kWh$_e$]
HYDROPOWER	0.01 - 0.04	0.02 - 0.04
OCEAN THERMAL ENERGY CONVESRION	0.17 - 0.22	0.07 - 0.10
WAVE POWER - small - large	≈ 0.09 -	0.04 - 0.07 0.10 - 0.17
TIDAL POWER	-	0.08 - 0.16
WIND POWER - small - large	0.4 - 0.06 0.06 - 0.09	0.025 - 0.04 0.04 ?
DIRECT SOLAR - PHOTOVOLTAIC (USA)	0.43	2000: 0.17 2030: 0.03 - 0.06
DIRECT SOLAR - SOLAR THERMAL	0.07 - 0.10	< 0.07
BIOMASS	0.07 - 0.085	0.03 - 0.05
RANGE OF ALL TECHNOLOGIES	0.01 - 0.43	0.02 - 0.17

LBST Ludwig-Bölkow-Systemtechnik GmbH /Wu 1993-05-10

Fig. 4.13 Electricity production costs for renewable energy sources.

4.4.2 OTEC

In the long term the electricity production costs from OTEC – depending very much on the site of application – will drop to about 0.12 US$/kWh (2010) and to approximately 0.08 US$/kWh (2030) from their present assumed level of approximately 0.20 US$/kWh (based on research plant status) [15] (see Fig. 4.13)

4.4.3 Wave power

For small units (a few W to some 100 kW) figures of 0.05–0.07 US$/kWh are realistic [46]. These figures are also estimated based on the Norwegian Norwave feasibility study undertaken for Indonesia [55]. For larger plants (MW range) figures of 0.12–0.20 US$/kWh seem reliably realizable, figures of 0.06–0.10 US$/kWh are disputed by proponents of the technology as feasible [46] (see Fig. 4.13).

Annotation: 'The latest investigations by the Fraunhofer Institute for Systems and Innovation Research, in Karlsruhe, have shown that the total average costs of electricity produced by wave power in the North Sea (shoreline and offshore stations) will range from 0.08 ECU/kWh to 0.14 ECU/kWh. As more and more wave power systems are built and experience gained, the cost could continue to fall. This shows that wave power as a source of hydrogen produced cleanly cannot compete with hydropower, but it would be better than exploiting wind power systems or photovoltaics. That is why the technical development in this field should be observed carefully. The successful penetration of cleanly produced hydrogen into the energy market presupposes large scale production. This also directs attention towards the generation of electricity by means which are as yet undeveloped. From this point of view wave power is an interesting option' [45].

4.4.4 Tidal power

The technology for tidal energy plant development has matured during the last 30 years, mainly due to improvements in barrage configuration and operation, in materials and construction methods (e.g. caissons) and in machinery. Therefore, cost reduction potentials for tidal electricity seem only moderate, and are estimated to be in the order of 10 to 20% from present levels. Machine efficiency and reliability is already high and thus the potential for further improvements seems limited [54].

The economic potential for the large tidal power schemes in planning state in Europe (see Fig. 4.13) with an assumed lifetime of 120 years lie in the order of equal or above 9 ¢US/kWh (at discount rates above 10%) or 18 ¢US/kWh (at discount rates above 15%). Potential development sites are under investigation in Argentina, Australia, Canada, India, Korea, Mexico, United Kingdom, United States and Russia, and amount to a total projected capacity of 87.5 GW_e and an energy yield of approximately 260 TWh/year [54].

4.4.5 Wind energy

In areas with an average annual wind speed of 8 to 9 m/s, large-sized turbines lead to generation costs of 7 to 10 ¢US/kWh (rates for consolidated technologies are 4 to 7 ¢US/kWh) and a cost reduction to 5 ¢US/kWh is presently considered achievable. In the long term (30 years) wind could see costs reduced to as low as 3 ¢US/kWh [16] (see Fig. 4.13).

4.4.6 Solar energy

The direct conversion efficiency (solar radiation/electricity) for the two most promising technologies – solar thermal parabolic trough electricity generating systems (SEGS) and photovoltaic electricity generating systems (PV) – presently are in the order of 30–38% for SEGS (south of 35 °N latitude) and between 5% (amorphous silicon), 17% (crystalline silicon) and above 25% (GaAs) for commercially available technologies. The annual system efficiencies for plants without storage are in the order of 10–15% for SEGS and between 4% and 10% for PV plants.

In gas hybrid operation (25%) energy production costs can be as low as 0.12 US$/kWh in California (1990) and finally can be reduced to about below 0.08 US$/kWh in the mid-1990s for SEGS [54] (see Fig. 4.13 and 4.14), whereas for utility scale PV the present costs under Californian insolation conditions have to be assumed realistically to be still in the order of 0.50 US$/kWh (1990) with the potential to drop to 0.20 US$/kWh (2000) and finally to about 0.04 to 0.07 US$/kWh (2030) [15] (see Fig. 4.13 and 4.15).

Besides parabolic trough collector systems, for direct solar electricity generation also central receiver systems (> 10 MW$_e$ to several 100 MW$_e$)

	PARABOLIC TROUGH			CENTRAL RECEIVER				DISH–STIRLING		
Timeframe	80 MW$_e$ LS-3 Present	80 MW$_e$ LS-4 1995–2000	200 MW$_e$ LS-4 2000–2005	100 MW$_e$ first plant 1995	200 MW$_e$ first plant 2005	200 MW$_e$ baseload 2005–2010	200 MW$_e$ advanced receiver 2005–2010	3MW$_e$/per year early remote market 1995–2000	30 MW$_e$/per year early utility market 2000–2005	300 MW$_e$/per year utility market 2005–2010
Capital cost range $/kW$_e$	3,500–2,800	3,000–2,400	2,400–2,000	4,000–3,000	3,000–2,225	3,500–2,900	2,500–1,800	5,000–3,000	3,500–2,000	2,000–1,250
Collector system typical cost $/m²	250	200	150	175–120	120–75	75	75	500–300	300–200	200–150
Annual solar-to-electric range[a]		13–17 percent		8–15 percent	10–16 percent		12–18 percent	16–24 percent	18–26 percent	20–28 percent
Enhanced load matching method	25 percent natural gas	25 percent natural gas	25 percent natural gas	Thermal storage	Thermal storage	Thermal storage	Thermal storage	Solar only	Solar only	Solar only
Solar capacity factor range	22–25 percent	18–26 percent	22–27 percent	25–40 percent	30–40 percent	55–63 percent	32–43 percent	16–22 percent	20–26 percent	22–28 percent
Annual O&M cost range ¢/kWh	2.5–1.8	2.4–1.6	2.0–1.3	1.9–1.3	1.2–0.8	0.8–0.5	1.2–0.8	5.0–2.5	3.0–2.0	2.5–1.5
Solar LEC range[b] ¢/kWh	16.7–11.8	17.2–9.8	11.7–7.9	16.1–8.0	10.1–5.8	6.5–4.6	8.2–4.5	32.8–14.6	18.6–8.8	10.6–5.5
Hybrid LEC range[b] ¢/kWh	13.0–9.3	13.5–7.9	9.3–6.5	–	–	–	–	–	–	–

The data in this table are compiled from several sources. The most comprehensive source is the U.S. DOE analysis performed for the National Energy Plan [3, 95–98]. The LEC calculations are based on a 6% real discount rate. They differ slightly from other LEC values given in this chapter that were calculated using different sets of economic assumptions.
a. Typical southwest US site.
b. Fixed charge rate: 7.8%.

Fig. 4.14 Levelized energy cost projections of solar thermal power plants [54].

	Units	1.8x10⁵ m² ~ 10 MW	4.2x10⁵ m² ~ 30 MW	1.1x10⁶ m² ~ 100 MW	3.7x10⁶ m² ~ 500 MW
Efficiency	%	7	9	11	15
Yield	%	80	80	85	90
Depreciation period	years	5	5	7	10
Capital investment	$ millions	18	32	70	185
Investment/capacity	$/W	2	1	0.7	0.4
Equipment	$/m²	20	15	0.7	5
Labor and fringe	$/m²	20	17	13	10
Materials	$/m²	30	25	25	20
Utilities	$/m²	10	8	7	6
Overhead	$/m²	10	7	5	3
Cost	$/m²	113	90	70	49
	$/W	1.6	1.0	0.64	0.33
PV system cost	$/W$_p$	3	1.9	1.3	0.76
Sunlight					
below average	$/kWh	0.29	0.18	0.12	0.07
average	$/kWh	0.24	0.15	0.10	0.06
desert	$/kWh	0.18	0.11	0.075	0.045

a. Plant capacities in MW assume stated module efficiencies and production yields, which vary for assumed maturity. Three shifts and 260 workdays are assumed at all production levels. The equipment costs drop rapidly in the larger plants because longer depreciation periods are assumed. Final module costs include yield losses (they are the total of the component costs divided by the yield). PV system cost assumes balance-of-system costs in $/m² of $100, $80, $70 and $65 for fixed flat-plates; a below-average location like Long Island, New York (1500 kWh/m² of sunlight each year), an average location like Kansas City, Missouri (about 1800 kWh/m²), and a desert location like Phoenix, Arizona (2400 kWh/m²); system efficiencies have been reduced 20 % to account for losses expected under typical field conditions.

Fig. 4.15 Possible evolution of PV-costs as plant size and technical maturity increase[a] [54].

and parabolic dish stirling generating units (5 to 25 kW$_e$, possibly larger in the future) are possible options. Central receiver systems are typically utility tailored and seem to have the potential for reaching or even undercutting the production costs of parabolic trough systems. Parabolic dish stirling systems are ideally suited for decentralized applications where they can compete with diesel systems already, but can also be integrated into farm plants. If produced in a large scale, electricity production costs can be reduced significantly but supposedly will not be able to compete with central solar or parabolic trough.

In Germany PV production costs are presently in the order of 0.75–1.0 ECU/kWh, whereas in Switzerland they are already closer to or

	CS[a]	BS[b]	CIG/STIG[c]	BIG/STIG[d]	CIG/ISTIG[c]	BIG/ISTIG[d]
Fuel[e]	$1.061 \cdot P_c$	$1.536 \cdot P_b$	$1.011 \cdot P_c$	$0.992 \cdot P_b$	$0.855 \cdot P_c$	$0.839 \cdot P_b$
Labour			0.30	0.20	0.28	0.19
Maintenance			0.42	0.32	0.33	0.24
Administration			0.14	0.10	0.12	0.09
Total fixed O&M	**0.35**	**0.80**	**0.86**	**0.62**	**0.73**	**0.52**
Water requirements			0.028	0.028	0.026	0.026
Catalysts/binder			0.018	–	0.016	–
Solids disposal			0.071	0.069	0.060	0.059
H_2SO_4 by-product credit			−0.273	–	−0.231	–
Total variable O&M	**0.59**	**0.50**	**−0.16**	**0.10**	**−0.13**	**0.09**
Capital[m]						
6% discount rate	1.71	2.27	1.66	1.32	1.32	1.03
12% discount rate	3.19	3.99	2.85	2.26	2.26	1.76
Total	$1.061 \cdot P_c +$	$1.536 \cdot P_b +$	$1.011 \cdot P_c +$	$0.992 \cdot P_b +$	$0.855 \cdot P_c +$	$0.839 \cdot P_b +$
6% discount rate	2.65	3.57	2.36	2.04	1.92	1.64
12% discount rate	4.13	5.29	3.55	2.98	2.86	2.37
	\multicolumn{6}{c}{Examples (total busbar cost):}
For P_c = $1.8 per GJ						
6% discount rate	4.56		4.18		3.46	
12% discount rate	6.04		5.37		4.40	
For P_b = $3.0 per GJ						
6% discount rate		8.18		5.02		4.16
12% discount rate		9.90		5.96		4.89

a. CS = a subcritical, coal-fired steam-electric plant (two 500 MW_e units) with flue gas desulfurization, east or west central US siting. EPRI estimates for heat rate (10.61 megajoules per kWh), overnight construction cost ($1217 per kW_e), other capital ($78 per kW_e), O&M costs ($23.1 per kilowatt per year fixed; $0.0059 per kilowatt variable), and the idealized plant construction time (5 years) [60]. Including AFDC, the total capital cost amounts to $1450 per kW_e ($1624 per kW_e) for a 6% (12%) discount rate.

b. BS = a 27.6 MW_e biomass-fired steam-electric plant. Based on an EPRI design for a 24 MW_e condensing/extraction cogeneration plant producing 20,430 kilograms per hour of steam at 11.2 bar for process [60]. Here it is assumed that this steam is instead condensed, thus producing an additional 3.6 MW_e; the heat rate is 15.36 megajoules per kWh (corresponding to steam conditions of 86 bar and 510 °C at the turbine inlet and a turbine efficiency of 80%). EPRI estimates for the over-night construction cost ($1693 per kW_e), other capital ($127 per kW_e), and idealized construction period (3 years). Including AFDC, the total capital cost amounts to $1,924 per kW_e ($2031 per kW_e) for a 6% (12%) discount rate.

c. CIG/STIG = a coal-integrated gasifier/steam-injected gas turbine and CIG/ISTIG = a coal-integrated gasifier/intercooled steam-injected gas turbine (see table 2).

d. BIG/STIG = a biomass-integrated gasifier/steam-injected gas turbine and BIG/ISTIG = a biomass-integrated gasifier/intercooled steam-injected gas turbine (see table 2).

e. P_c = coal price, and P_b = biomass price, in $ per gigajoule (HHV basis).

Fig. 4.16 Busbar costs for alternative power technologies (cents per kWh) [54].

below 0.75 ECU/kWh. In detailed studies [22] it can be shown that with existing mass produced multi-crystalline PV technology electricity production costs as low as 0.35 ECU/kWh can be achieved. In Japan production costs for grid-connected roof top integrated PV applications of about 350 ¥/kW$_{p,e}$ shall be achieved in the year 2000.

4.4.7 Bioenergy

Cost comparisons of wood and straw with conventional fuels harmonized on the basis of the utilizable thermal energy content show that these biofuels can be cost competitive with coal, oil or gas. Depending on the regional availability of the biomass, costs of between 20% to 80% of the costs from fossil fuels can be achieved, i.e. between 4.5 DM/GJ and 16 DM/GJ [53]. Electricity prices from wood-fired plants are in the range of 0.04–0.05 US$/kWh$_e$.

For biomass produced from energy crops in energy plantations costs of approximately 100 DM/t = 3.25 US$/GJ are assumed for Germany (for miscanthus) [52] and about 2.5 US$/GJ for the USA (for wood) [52]. To be competitive with conventional fossil fuels costs of energy crops have to drop to approximately 2 US$/GJ [15].

Achievable electricity costs from biomass-integrated gasifier/steam-injected gas turbine plants respectively from biomass-integrated gasifier/intercooled steam-injected gas turbine plants on the basis of a biomass price of between 1.8 and 3.0 US$/GJ are estimated to be in the range of 0.04–0.06 US$/kWh$_e$ [54] (see Fig. 4.13 and 4.16).

Studies [52] have shown that hydrogen production from wood, peat, straw, dry residues and miscanthus senensis in plants of at least 300 MW$_{th}$ (capacity approximately 1500 t/d) operated approximately 8000 h/yr is possible at costs of 0.27 DM/Nm^3. Conversion efficiencies of between 57% (miscanthus → hydrogen) and 69% (wood → hydrogen) can be achieved [52].

4.5 CONCLUSIONS AND RECOMMENDATIONS

4.5.1 Total potential for renewable energy sources

The remaining potential for renewable electricity production as depicted in Fig. 4.8 is about 19 times as high as present world electricity production. This estimation for the remaining potential already excludes potentials which seem not to be utilizable due to yet known environmental restrictions

68   CLEAN ENERGY SOURCES

(e.g. for hydropower, wind energy). Even if only 10% of this potential were used for electricity production it could cover even growing world needs far into the 21st century.

This leads to the conclusion that area requirement and potentials are not the limiting factors for large-scale renewable energy use, as also societal and environmental restrictions need not to be the limiting factors.

4.5.2 Electricity generation costs

Presently two electricity generating concepts clearly can be identified as the most economic ones, i.e. hydropower (0.01–0.04 ECU/kWh$_e$) and wind energy conversion (0.04–0.06 ECU/kWh$_e$). By far the most expensive technology is photovoltaics (0.43 ECU/kWh$_e$).

Looking into future technologies and applications many more concepts are found in similar range of generating costs (0.02–0.06 ECU/kWh$_e$) which are hydropower, small wave power, wind power, solar thermal energy conversion, photovoltaics and biomass (see Fig. 4.13).

This leads to the conclusion that in the long run costs are not the only limiting factor for large-scale renewable energy use.

4.5.3 Timely availability of the investigated concepts

Status: Among the presently proven technologies for renewable energy/electricity production count hydropower for hydro-electricity generation (up to several GW$_e$), tidal power electricity generation (up to 240 MW$_e$), small and medium-size wind turbines for electricity generation (from a few kW$_e$ up to some 700 kW$_e$), solar thermal flat plate collectors for hot water and process steam production (from a few kW$_{th}$ to several MW$_{th}$), solar thermal parabolic trough power plants (up to units of 80 MW$_e$) and parabolic dish stirling power plants (up to units of 25 kW$_e$) for electricity generation, small stand-alone or grid-connected photovoltaic plants for electricity production (some 100 W$_e$ to some 100 kW$_e$, only a few plants in the X MW$_e$ range) and biomass for energy production via incineration of biomass for process steam/heat and/or electricity production (in the range of some MW$_e$ to some X0 MW$_e$). OTEC and wave power have not yet emerged from the prototype state of development.

Judged from present state of technology and economics, by far the cheapest and maturest technology is hydropower. Small wind power units already come very close to hydropower and are followed by biomass and by solar thermal parabolic trough plants. OTEC, wave power and photovoltaic are either not yet mature or still too expensive.

Forecast: For the future, the timely availability of the different concepts only partly depends on the technology, rather more on economic and environmental constraints. Hydropower still will have a large potential for renewable electricity production in the multi-MW-scale up to the GW-scale. Wind energy will be very close in economics to hydropower for small/ medium-scale turbines. Siting constraints in densely populated areas will require large-scale turbines (1 to 3 MW_e) with still poorer economics. Electricity from biomass with more advanced technologies (integrated gasifier/intercooled steam-injected gas turbine plants) will achieve economics very close to hydropower and wind energy.

At about the same time – i.e. approximately within a decade from now – solar thermal parabolic trough power plants will reach comparable generating costs and can be employed on a large scale (units of up to 160 MW_e and farm configurations of several times this capacity). Also small-scale wave power units might achieve similar economics at the same time.

Before photovoltaics will achieve the same generating costs as hydro, wind, biomass and solar thermal presumably one or two more decades will have to elapse. According to present knowledge, OTEC and large wave power seem not to have the potential to achieve the same low generating costs, along with tidal power plants.

Conclusion: These considerations point out that the readily available large-scale renewable electricity production technology is hydropower. There is still a vast potential for its extended low cost use if hydropower developments are carried out in a way compatible with ecological and societal requirements.

With regard to technology and achievable electricity generating costs, wind energy conversion, solar thermal parabolic trough plants and biomass plants will become available commercially for large-scale electricity production within the next decade. Biomass is the only source which can also be converted in the large-scale into hydrogen directly via gasification.

Advanced tidal power plants will need one additional decade to be realized at large scale (2010). OTEC may have a huge potential for electrolytic hydrogen production (due to its remoteness in siting), but it is not yet sure if it will be realized economically, therefore it is estimated not be available before the year 2020.

By far the highest cost reduction potential as well as vast potential for application (due to its versatility in almost latitude-independent application) is photovoltaic power. If no major efforts are undertaken for its large-scale introduction, economic application for bulk electricity production is assumed for 2025. If increased efforts are undertaken to advance technology and costs, this field of application could be available one or two decades earlier.

This leads to the conclusion that in the long run, neither technical availability nor economic viability have to be the limiting factor for large-scale renewable energy use.

Remarks: For all large-scale applications of renewable energy technologies (due to their low energy density and thus high area requirements) environmental considerations (setbacks as well as benefits) need to be noted as the most important ones, besides economics.

For a successful introduction of renewable energy sources and hydrogen we first of all need the perception and the conviction for a sustainable and environmentally compatible, economic energy system. In order to internalize external and social costs related to energy production from conventional sources, to improve overall energy efficiency, to give renewable energies and hydrogen a chance and to promote their widespread application it will be necessary to artificially increase relative energy prices to come to a *real cost pricing*.

The very often cited argument that Germany would lose its international competitiveness seems weak, since in the average German industrial product costs the energy cost share contained amounts to no more than 4%.

The money recovered by the government through energy taxation can be used to improve/support the introduction of a clean transport and energy infrastructure and to lower taxes and duties on human labour (e.g. income tax) and on private investment.

In the long run, internalization of external effects will only work if a sufficient political majority and a broad acceptability by the population can be achieved. This will also mean that a broad consensus about a future concept for energy generation and utilization (i.e. the energy system) will have to be achieved in society.

In the wake of rapidly growing world population and changing climate patterns fast, and preferably harmonized, action is required worldwide for the improvement of energy efficiency, the optimization of appropriate supply (infra) structures and the provision of clean and environmentally compatible energy sources. Although it still may take some time to come to international conventions, national measures to improve the situation should not be postponed with the common argument that an international harmonization is still missing and that due to economic reasons one cannot take the first step. Some countries which invested in environmental protection technologies first, were also the biggest players in these markets later on (e.g. Japan and Germany). Maybe this can serve as an example for an initiative on large-scale use of renewable energy and hydrogen technologies.

4.6 REFERENCES

1 *Energieversorgung der Zukunft*, von J. Nitsch und J. Luther, Springer-Verlag, 1990.

2 'Weltweite Nutzung der Wasserkraft – Wasserstoff für Europa?', von H. Blind, in *Energiewirtschaftliche Tagesfragen*, Heft 6, Juni 1988.

3 Enquête Kommission des Deutschen Bundestages 'Technikfolgen-Abschätzung und -Bewertung: Aufbaustrategien für eine solare Wasserstoffwirtschaft', Endbericht und Materialienbände I-V, Berlin, Bonn, Köln, Ottobrunn, Stuttgart, März – Juni 1990.

4 Seasonal Storage of Hydrogen in liquid organic Hydrides: Description of the second prototype Vehicle by Grünenfelder and Schucan, in *Int. J. Hydrogen Energy*, Vol. 14, No. 8, pp. 579–586, 1989.

5 Wasserstoff aus Wasserkraft – die unerschöpfliche heimische Energiequelle, Hans Breuer, in *Energie Aktuell*, Heft 14, 8. Jg., 1984.

6 Entscheidungen für eine langfristige Energiepolitik; von Ludwig Bölkow; in *Peutinger-Collegium*, München, März 1982.

7 *Wasserstoff als Energieträger – Technik, Systeme*, Wirtschaft, von C.J. Winter und J. Nitsch, bei Springer-Verlag, 1986.

8 Replacement of Fossil Fuels by Hydrogen, by R. Dahlberg; in *Int. J. Hydrogen Energy*, Vol. 7, No. 2, 1982, pp. 121.

9 *Energy Options*, J.O.M. Bockris; Taylor and Francis, London 1980.

10 *Hydrogen and the Revolution in Amorphous Silicon Solar Cell Technology*, by J.M. Ogden and R.H. Williams, Center for Energy and Environmental Studies Princeton University, PU/CEES Report No. 231, February 15, 1989.

11 'Unerschöpfliche saubere Wasser- und Energiequellen in Grönland', von C.F. Kollbrunner/H. Stauber, bei Verlag Leemann, Zürich, 1973.

12 '*Energie und Wasser aus Grönland*', von Curt F. Kollbrunner, bei Verlag Leemann, Zürich 1976.

13 *International Water Power & Dam Construction*, Vol. 41, No. 9, September 1989, pp. 32.

14 *Hydrogen Production and Exportation in Québec – a Techno-economic Analysis*, by A. Haurie and R. Loulou, Group d'Etudes et de Recherche en Analyse des Décisions (G.E.R.A.D.), Work Package No. 242, Euro-Québec Hydro-Hydrogen Pilot Project, January 1991.

15 *The Potential of Renewable Energy – An Interlaboratory White Paper*, prepared for the Office of Policy, Planning and Analysis, U.S. Department of Energy under Contr. No. DE-AC02-83CH10093 by Idaho National Engineering Laboratory (INEL); Los Alamos National Laboratory; Oak Ridge National Laboratory (ORNL); Sandia National Laboratories; Solar Energy Research Institute (SERI); Solar Energy Research Institute SERI, Golden, Colorado March 1990.

16 *Macro Analysis*, Work Package No. 210, Euro-Québec Hydro-Hydrogen Pilot Project, associated contractor to contract no. 3723-89-05 PC ISP D given by the European Atomic Energy Community, Fraunhofer-Institut für Systemtechnik und Innovationsforschung FHG-ISI, December 1990.

17 *Wind Energy, Potential of Wind Energy in the European Community*. An Assessment Study, by H. Selzer, D. Reidel Publ. Company, 1986.

18 Anwendung und Weiterentwicklung umweltverträglicher regenerativer Energietechnologien in Marokko – Konzept für eine Kooperation zwischen dem Königreich Marokko und dem Freistaat Bayern; a short study prepared for the Bavarian Minstries for State Planning and Environmental affairs and for Economy and Transport; in cooperation with the Centre National de Coordination et de Planification de la Recherche Scientifique et Technique, Rabat; carried out by Ludwig-Bölkow-Systemtechnik GmbH; Munich, April 1991.

19 Calculated costs for the 100th Swedish wind turbine system, *IEA Economic Aspects of Wind Turbine LS-WECS Expert Meeting on Costing*, Petten, 30–31 May 1985; by A.-K. Larssen, Daten aus: Vindkraft, Statens energiverk 1985: 1, Stockholm.

20 Various internal considerations and calculations performed in the field of solar energy generation and hydrogen production, Ludwig-Bölkow-Systemtechnik GmbH, 1982–1991, [not public].

21 Perspektiven der Photovoltaik, von K. Hassmann, W. Keller, D. Stahl, Siemens-Erlangen, in BWK Bd. 43 (1991) Nr.3 – März, pp. 103–112.

22 Kostendegression Photovoltaik, Stufe 1: Fertigung multikristalliner Solarzellen und ihr Einsatz im Kraftwerksbereich, by J. Schindler and D. Strese, research report by Ludwig-Bölkow-Systemtechnik GmbH, prepared for the German Ministry of Research and Technology, Ottobrunn 1988.

23 Einsatzmöglichkeiten der Photovoltaik zur Entlastung des Energiemarktes, by D. Strese, Ludwig-Bölkow-Systemtechnik GmbH, Ottobrunn 1989.

24 Thin Film Photovoltaics, paper prepared for the *24th Intersocoety Energy Conversion Engineering Conference*, by K. Zweibel and H.S. Ullal, Washington D.C. 6–11 August 1989, Solar Energy Research Institute, May 1989.

25 *Harnessing Solar Power – the Photovoltaic Challenge*, by Ken Zweibel, Plenum Press New York and London, 1990.

26 *Energy System Emissions and Materiel Requirements*, prepared for: The Deputy Assistant Secretary for Renewable Energy – U.S. Department of Energy, Washington, D.C., by Meridian Corporation, Alexandria, Virginia, February, 1989.

27 *Euro-Québec Hydro-Hydrogen Pilot Project Phase II Feasibility Study* work carried out under the contracts no. 3549-88-12 PD ISP D and 3723-89-05 P-C ISP D given by the European Atomic Energy Community, carried out by Ludwig-Bölkow-Stiftung, Ottobrunn, 1991, (not yet published – March 1991).

28 The Euro-Québec Hydro-Hydrogen Pilot Project EQHHPP, R. Wurster, paper

presented at the *8th World Hydrogen Energy Conference*, Hawaii, July 22–27, 1990, Pergamon Press 1990.

29 Solarwasserstoffprojekte – Pilot- und Demonstrationsprojekte, R. Wurster, in Sonnenenergie 6/90.

30 HNEI Wind-Hydrogen Programm, D. Neill *et al.*, paper presented at the *8th World Hydrogen Energy Conference*, Hawaii, July 22–27, 1990.

31 Optimierung eines Energieversorgungssystems auf der Basis von Windenergie und Wasserstoff, von Schulien/Steinmetz, in 7. Intern. Sonnenforum, DGS, Frankfurt 9.-12. Okt. 1990.

32 Hydrogen Energy Storage for an Autonomous Renewable Energy System – Analysis of Experimental Results, abstract of a paper to be presented at the *ISES conference* in Denver, Fall 1991, by A. Haas, J. Luther, F. Trieb, University of Oldenburg.

33 Realisierung des Solar-Wasserstoff-Projekts Neunburg vorm Wald, Axel Szyszka, SWB GmbH, München, in Elektritzitätswirtschaft, Jg. 89 (1990), Heft 5.

34 Hysolar – Solar Hydrogen Energy/A German – Saudi Arabian Partnership, DLR, KACST, 1989 (and personal communications H. Steeb).

35 Das Energieautarke Solarhaus, W. Stahl, A. Goetzberger, Fhg-ISE Freiburg, in 7. Intern. Sonnenforum, DGS, Frankfurt 9.–12. Okt. 1990.

36 Das energieautarke Solarhaus – Entwicklung des Fraunhofer-Instituts für Solare Energiesysteme, by W. Stahl and A. Goetzberger, in Sonnenenergie 6/90, December 1990.

37 Europas einziges 'Wasserstoff-Wohnhaus' – im Emmental, Rudolf Weber, 1990.

38 Deutsche Airbus GmbH (MBB Transport Aircraft Group); Pilotprojekt Airbus mit Wasserstoffantrieb 'Einsatz von LH_2 in der Luftfahrt', Hamburg 1989.

39 Electric Power in Canada 1988, Minister of Supply and Services Canada, Ottawa 1989.

40 *Energy in Québec* 1988 Edition, Gouvernement du Québec, Ministère de l'Énergie et des Ressources, Québec 4/1988.

41 Appraisal of the Hydropower Potential in Iceland, Landsvirkjun, Reykjavik 1986.

42 *Energy in a Finite World – a Global Systems Analysis*; by W. Häfele; Ballinger Publishing, Cambridge, Mass.; 1981.

43 The Contribution of Hydrogen in the Development of Renewable Energy Sources; by J. Nitsch, H. Klaiß, J. Meyer of DLR, Stuttgart; paper presented *at the 8th World Hydrogen Energy Conference*, Hawaii, July 22–27, 1990, Pergamon Press 1990.

44 *Greenpeace, Energy Without Oil – The Technical and Economic Feasibility of Phasing Out Global Oil Use*, Energy Policy and Research Unit, Greenpeace

International, Keizersgracht 176, NL-1016 DW Amsterdam, The Netherlands, January 1993.

45 *Hydrogen as Energy Vector – An Assessment on Future Hydrogen Markets*, Fraunhofer Institut für Systemtechnik und Innovationsforschung – ISI, EQHHPP Supplementary Task Program work package WP 350 [contract no. 3723-89-05 PC ISP D by the European Atomic Energy Community].

46 1989 Survey of Energy Resources, *World Energy Conference*, Montreal 1989.

47 Energie und Klima, herausgegeben von der Enquete-Kommission 'Vorsorge zum Schutz der Erdatmosphäre' des Deutschen Bundestages, Band 3: Erneuerbare Energien, Economica Verlag, Bonn, und Verlag C.F. Müller, Karlsruhe, 1990.

48 Fuels from the Oceans via Ocean Thermal Energy Conversion (OTEC) by W.H. Avery, Proceedings of Ocean Space Utilization Conference, Tokyo, Japan, 1985.

49 *Weltressourcen*, World Resources Institute and International Institute for Environment and Development, ecomed, 1991.

50 *Hydro Power: The Design, Use and Function of Hydromechanical, Hydraulic and Electrical Equipment*, by J. Raabe, VDI-Verlag, Düsseldorf, 1985.

51 Möglichkeiten und Potentiale zur Nutzung und Produktion von Pflanzen als Träger gespeicherter Sonnenenergie – eine globale Betrachtung, A. Strehler, Landtechnik Weihenstephan, 7. Intl. Sonnenforum der DGS, Frankfurt 1990.

52 Bedeutung, Einsatzbereiche und technisch-ökonomische Entwicklungspotential von Wasserstoffnutzungstechniken – Band II: Indirekt energetische Nutzung; Einsatzbereiche und Potential der Nutzungstechniken, ZSW-DLR-LBST, Stuttgart – Ottobrunn, März 1992.

53 Wärme aus Stroh und Holz, A. Strehler, Bayerische Landesanstalt für Landtechnik, Freising, 1980.

54 *Renewable Energy – Sources for Fuels and Electricity*, Editors: T.B. Johansson, H. Kelly, A.K.N. Reddy, R.H. Williams, Exec. Editor: Laurie Burnham, Island Press, Washington D.C. – Covelo, California, 1993.

55 *NORWAVE Wave Power for Island Communities – a Status Report*, NORWAVE A.S., Oslo, Norway, 1988.

56 Systemvergleich und Potential von solarthermischen Anlagen im Mittelmeerraum [Systems Comparison and Potential of Solar Thermal Installations in the Mediterranean Area], G. Hille, H. Klaiß, J. Meyer, F. Staiß, P. Wehowsky, DLR – Interatom – SBP – ZSW, Stuttgart, Juli 1991.

57 Vertical axis wind turbine with integrated magnetic generator, G. Heidelberg, J. Kroemer; Heidelberg Motor GmbH, Symposium "Windenergie Bremen '90", 27 June 1990.

5
Technology for Gaseous Hydrogen Production

H. Buchner

Daimler-Benz Central Research, Stuttgart, Germany

5.1 CURRENT STATE OF H_2 PRODUCTION

Although the central production and distribution of hydrogen in gaseous or liquid form is possible on an industrial scale, the financial investment required is, in general, extremely high. A wide-ranging urban and rural hydrogen infrastructure for the exclusive use of vehicles is therefore not feasible in the foreseeable future for economic reasons.

The solution to the problem of how to supply vehicles with hydrogen on a local basis is to be found in the use of gas and electricity, i.e. using already existing energy carriers and their infrastructures. Up to now, these energy sources have been used almost exclusively for stationary energy supplies rather than for vehicular operation.

Most of today's hydrogen is produced by cracking natural gas in central installations with power capacities in excess of 50 MW. The process: steam reforming (H_2O and CH_4) with subsequent conversion (CO and H_2O). Hydrogen costs are in the order of 0.60 DM per litre gasoline equivalent.

Hydrogen is also produced as a by-product of the chemical industry (refineries, Hüls Co. and others). A hydrogen pipeline network covers the entire Ruhr conurbation and extends as far as Düsseldorf and Cologne.

Production of hydrogen in large-scale electrolysis units using cheap electricity results in hydrogen cost between 1.00 and 1.50 DM per litre gasoline equivalent. The oxygen produced by the electrolysis (8 t O_2/t H_2) can also be used to reduce the hydrogen cost.

5.2 LIQUID HYDROGEN

LH_2 can only be economically produced in central installations. The current delivery price of liquid hydrogen (from natural gas or chemical H_2) is approximately 4.00 DM per litre gasoline equivalent (untaxed!). LH_2 can only be transported using expensive heat-insulated tankers. These tankers contain the energy equivalent of approximately a quarter of the same tank filled with gasoline (fourfold tanker capacity and fourfold storage capacity compared to gasoline). Storage over weeks is associated with high evaporation losses (2% LH_2 loss per day from a vehicle tank). The supply of aircraft therefore requires the provision of a liquefaction installation at the airport (with GH_2 from the vehicle supply infrastructure via gas pipelines).

5.3 INTEGRATION OF HYDROGEN SUPPLY FOR VEHICLES AND PLANES INTO THE ELECTRIC POWER GRID

5.3.1 Water-electrolyzer as a power regulating component in the electrical grid

The following idea may result in the utilization of an electrolyzer as a regulating component in an electrical grid. Today's utility power supply is based on a momentary energy balance between generation and demand for electric power. The system is fed from a continuous use of fuel energy in power plants. Variations in demand are compensated for by regulating processes in the power plants. Some generating power has to be kept on standby in order to compensate for lack of generating power. The standby power is not available for the production process and therefore its cost has to be carried by the overall system.

The use of electrolyzers as regulators is shown in Fig. 5.1. The electrolyzer on the demand side of the system continuously generates hydrogen and oxygen as a commercial product of the utility. The regulating capability of the electrolyzer and its availability has such a quality that it may form a vital component of the overall generating system.

In case of sudden lack on the supply side the electrolyzer on the demand side may be switched off from full load in order to compensate for the lacking generating power. Also continuous power changes of the electrolyzer are made anticyclic to the demand changes at the grid. In this way the electrolyzer power is increased by the same amount as the remaining load decreases and vice versa. It is important for this proposed process that the

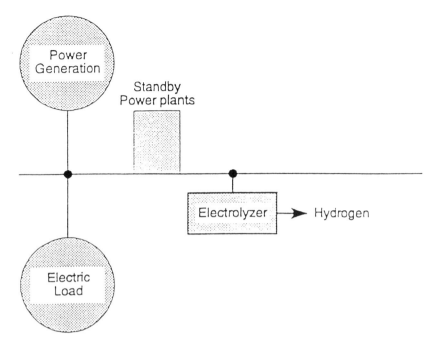

Fig. 5.1 Electrolyzer as regulating component in an electric power supply system.

electrolyzer be integrated as a grid component technically and also economically. This means that there must be high reliability inherent in the technology and a reliable long-term sale of the gases produced.

Figure 5.2 shows the variation of power generated by a power plant operating in regulating mode on an electrical utility grid. The main power value is set to 97.5% of rated power, the range of variation being +/− 2.5%. The marked areas in Fig. 5.2 correspond to amounts of electric energy, which is not generated because of the regulating function of the plant. This energy could be used to feed an electrolyzer. This results in a full load operation of the power plant. The cost for the electrical energy feeding the electrolyzer in this way may be counted low, e.g. fuel cost of the thermal power plant is in the order of 6 DPf/kWh.

On the basis of today's market price for hydrogen, the system may be economically successful depending on the value of the advantages encountered (e.g. avoidance of losses at the regulating valves of the steam turbine or prevention of air polution).

The amount of power required for grid regulation is about +/− 50 MW for utilities as Badenwerk AG or Hamburgische Elektrizitäts-Werke AG. In Germany a potential of about 1000 MW may be available for electrolyzers.

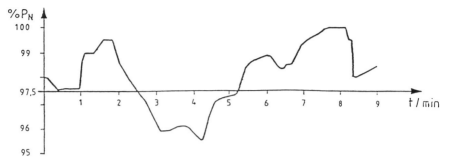

Fig. 5.2 Power output of a power plant in regulating mode.

The use of an electrolyzer as regulating load for an electrical power grid would be the first application of this kind.

5.3.2 Organization

Utilizing an electrolyzer as power regulation in a utility grid requires operation of the complete system by the utility on a long-term basis. This is also valid in cases where the electrolyzer is installed close to a hydrogen supply station at the city-grid. Because of the continuous operation of the regulating electrolyzer also a permanent sink for the gases has to be organized.

5.3.3 Hydrogen fuelling station with grid connection

A fuelling station for hydrogen-driven vehicles is planned in municipal areas (Karlsruhe, Hamburg). This hydrogen base consists of an electrolyzer with utility-controlled grid connection, hydrogen storage tank and a fuelling-tap.

The electrical power of for instance 2.5 MW is fed directly from a 110/10 or 20 kV substation. The hydrogen storage tank acts as a buffer between production and demand of hydrogen. In case of surplus production a connection to the natural city gas grid could act as a hydrogen sink (less than 5% H_2 in NG).

The oxygen produced in the electrolysis may be sold or let off into the air. The market value of the about 5000 Nm^3 O_2 gas produced per day is about 7.500 DM.

At continous full load the 2.5 MW electrolyzer produces 15.000 Nm^3 H_2 gas. In power grid regulation mode the production is about 8000 to 10.000 Nm^3, equivalent to about 3.000 l gasoline in 24 hours. This would fuel about 30 busses or 300 cars under city-traffic conditions. The space requirement of the electrolyzer including grid connection and storage tank would be about 50 to 100 m². The H_2 tapping unit required additionally to this area depends on whether buses are to be fuelled.

The hydrogen-fuelled fleet of vehicles should preferably run in the city. Considering service cars of municipalities (police, medical and social services) there will be a lack of fuel consumption during weekends and holidays. In order to overcome these periods, a continuous fuelling system for slow parallel operation on many vehicles should be developed.

5.4 DECOMPOSITION OF NATURAL GAS INTO HYDROGEN AND PURE CARBON

Kvaerner Engineering a.s. is presently developing a process which yields competitive and pollution-free pure carbon and hydrogen. With natural gas and electric power as input, the natural gas decomposes into pure carbon and hydrogen at high temperature (Fig. 5.3).

The process is one of the results of technological development performed

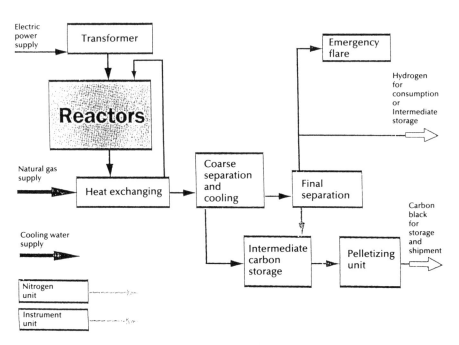

Fig. 5.3 CO_2-free decomposition of natural gas.

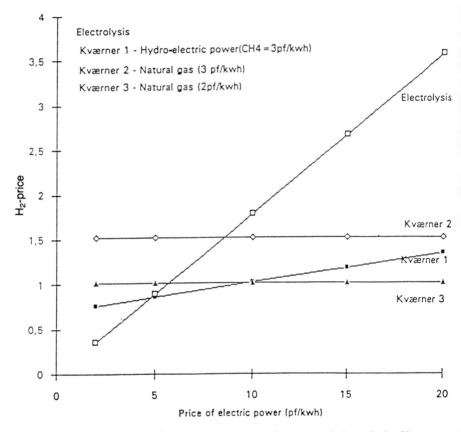

Fig. 5.4 H_2-price as function of electrical power for electrolysis and the Kvaerner process.

by Kvaerner Engineering and currently undergoing final verification and optimization before it is launched into the market. The following description will therefore give a brief outline of the process.

The big advantage of the process is that the decomposition of the natural gas takes place without any discharge of CO_2 or other 'greenhouse' gases harmful to the environment. One hundred percent exploitation of the carbon and hydrogen in the natural gas is achieved.

Yet another advantage is that hydrogen from the process can be utilized to generate electricity for the process and thereby avoid use of electric power generated by combustion of coal or hydrocarbons, thus avoiding global environmental emission. With an efficiency of 50% in electric power generation from hydrogen, less than 50% of the produced hydrogen is consumed in the process. With the development of more efficient systems, i.e. fuel cells, these figures could be improved considerably.

At present a pilot plant with an approximate capacity of two tonnes of hydrogen per day is undergoing testing and optimization. Experience and cost data from this test plant has been the basis for this estimate and should therefore give a very good indication of the potential in the process. The cost estimate is further based upon an optimization of the process for hydrogen generation and less attention is given to the carbon and carbon quality (i.e. particle size, surface, structure etc.). The estimate, therefore, is based on a relatively low price and value of the produced carbon (Fig. 5.4).

Furthermore, it is assumed that the hydrogen will be stored in pressurized tanks and therefore does not need to be of a high purity quality but could be accepted with traces of hydrocarbons. If high purity quality is needed the cost of a purificative unit needs to be added to the described plant and to the cost estimate.

The Kvaerner process is very flexible with regard to capacity utilization, because the main process is developed as a modular concept consisting of a number of reactors connected to a utility system. Both the size and the capacity of each reactor can be varied to a great extent.

The disadvantage might be that for very small capacities the utility systems could be expensive and make the process less competitive. On the other hand, this gives an advantage when it comes to increasing the capacity of a production plant.

The process has also the necessary flexibility with regard to shorter stops and restarts. For shorter shut-down periods the process can be restarted within seconds. At longer stops the start up flexibility can be retained by applying some energy to maintain a sufficient start up temperature of the reactors.

5.5 CONCEPT OF HYDROGEN PRODUCTION IN DECENTRALIZED INSTALLATIONS

The problems associated with an H_2 infrastructure for vehicles could be solved by decentralization. Gaseous hydrogen supplies for vehicles could be made available in conurbations in the short term, becoming extensively available in the medium to long term.

The production of hydrogen could be carried out directly at gasoline stations in electrolysis units or with gas fractionators (PSA) when, as has been demonstrated in Berlin, an available gas mixture contains hydrogen (from hydrogasification, natural gas reforming).

Hydrogen could even be produced for individual supply by means of small-scale electrolysis units (state-of-the-art). This, however, leads to the highest fuel costs of 2.50–3.00 DM per litre gasoline equivalent (untaxed!).

5.5.1 Hydrogen supply concept for households

In order to realize the benefits of hydrogen-fuelled, pollution-free automobiles, a hydrogen supply infrastructure must be established. The US company Hamilton Standard has designed a small SPE hydrogen generator based upon the principle of water electrolysis. SPE water electrolysis is a mature technology developed by Hamilton Standard for aerospace and defense applications.

SPE electrolyzers are unique in that there is no liquid electrolyte to present problems of leakage, corrosion or possible hazard in the event of an accident or spill. The solid polymer electrolyte is a solid sheet of plastic, similar to Teflon perfluorcarbon, which has been sulfonated to give it the capability to conduct hydrogen ions. Being a solid material, it provides a rugged barrier between the hydrogen and oxygen product gases, while providing a high electrical efficiency.

The size and weight of the unit is comparable to a portable air compressor. The only inputs to the generator are household tap water and electricity. A hydrogen supply line is connected to the automobile to charge the metal hydride tank. The SPE hydrogen generator can also be installed in the automobile offering the advantage of integrating the generator with the storage tank. The consumer simply fills the water reservoir and makes an electrical connection without having to connect the hydrogen supply line. Whether located in the garage or in the automobile, the system operates automatically and will shut off when the hydrogen tank is filled.

It is anticipated that hydrogen filling will occur overnight in the vehicle garage taking advantage of off-peak electrical demand periods. However, if the SPE hydrogen generator is built into the automobile, hydrogen charging can be accomplished anywhere electrical power is available. In either case, power demands are well within household service capabilities.

5.5.2 Energy balance

Producing GH_2 requires more energy than producing gasoline/kerosene from petroleum. The excess energy required is given below for different sources:

- natural gas: +28%
- coal: +46%

- electrolysis using current mix as in Germany: +325%
- electrolysis with hydroelectric power: +46%

5.5.3 CO₂ balance

Dependent on the primary energy used, hydrogen generation has the following CO_2 balances (rough estimate):

- Lower CO_2 emissions as compared to gasoline operation are caused by hydrogen generation from:
  - water power;
  - wind power, photovoltaic effect (not yet economically viable for H_2 generation);
  - nuclear energy;
  - current mix with $> 80\%$ CO_2-free primary energy (France, Norway, Canada, Austria, Switzerland and California);
  - natural gas steam reforming (the leakage of natural gas lines must also be taken into consideration).

  New procedures even allow CO_2-free generation of hydrogen from natural gas with carbon as a by-product for further industrial use (tyres etc.).

- Higher CO_2 emissions as compared to gasoline operation are caused by hydrogen generation from:
  - coal gasification;
  - conversion of coal into electric energy;
  - world current mix ($< 20\%$ CO_2-free primary energy carriers);
  - German current mix ($< 40\%$ CO_2-free primary energy carriers).

- LH_2 is generated from gaseous hydrogen when electric energy is added. The primary energy required for liquefaction corresponds roughly to the calorific value of gaseous hydrogen. This means particularly high CO_2 emissions (exception: use of water power, nuclear energy to generate LH_2).

- If due to limited oil reserves synthetic gasoline (made from coal and hydrogen) has to be increasingly used in the future, this will lead to higher CO_2 emissions than when using GH_2 and LH_2).

5.5.4 Cost of hydrogen

The cost of hydrogen is closely linked to the energy cost for natural gas and electric power.

The production cost of GH_2 generally varies between 0.60–1.50 DM/l gasoline equivalent. $3\,m^3$ of H_2 (= 1 litre gasoline equivalent), approximately 2 kg of oxygen and waste heat of approximately 1 litre heating oil equivalent are developed in electrolytic generation of hydrogen. Using the waste heat for heating service water or the oxygen in environmental technology (e.g. improving the quality of river water) reduces the cost of hydrogen by 30–50%.

The commercial price of LH_2 (1992) is in the order of 3.00–4.00 DM/l gasoline equivalent.

6
Technology for Cryofuel Production

P. Pelloux-Gervais

AIR LIQUIDE, Sassenage, France

6.1 PRODUCTION PROCESSES USED TODAY

To supply the Ariane 5 space project, L'Air Liquide has built its more recent liquid hydrogen plant in Kourou, French Guiana. Its production is about 2.5 t/d.

The main European plants of liquid hydrogen are:

- L'Air Liquide Waziers, France: 10.5 t/d
- Linde Ingolstadt, Germany: 4.4 t/d
- Air Products Rozenburg, Netherlands: 5 t/d

6.1.1 Gaseous hydrogen sources

Considering the high cost of electrical energy in French Guiana and also in Europe, water electrolysis was not selected to supply gaseous hydrogen. So the hydrogen source is the reforming of methanol. A chemical endothermic reaction products carbon dioxide and hydrogen from methanol and steam in presence of a catalyst.

For a similar reaction, merchant butane, naphta, kerosene or propane in Guiana and also natural gas in Europe could be used (there is no natural gas in Guiana). The choice results from the following criteria: supply safety, plant operation and level of investment or operation cost. This choice depends on local considerations.

In the other Air Liquide's plant in Europe, Waziers, the hydrogen source is an ammonia plant or an electrolyzer. We can also use as Linde at

Ingolstadt near München a hydrogen-rich raw gas of a refinery which is transported through a gas pipeline after purification. The Waziers plant is also linked with the hydrogen pipe network of Belgium and the north of France.

6.1.2 Purifying gaseous hydrogen

The liquefaction of hydrogen at −253 °C (or 20 K) first requires the removal of all impurities (water, oxygen, nitrogen, carbon dioxide etc.) which are solid at such a temperature. Only helium is gaseous at 20 K. The purification takes place in two stages:

- Hot purification with pressure swing adsorption (PSA):

    hydrogen from steam reforming is sent into a PSA which retains impurities and lets hydrogen pass first. The hydrogen leaving the PSA unit contains only a few ppm of H_2O, CO, CO_2, CH_4 etc.

- Cold purification:

    When the hydrogen is cooled to the temperature of liquid nitrogen (−196 °C), it is purified with activated charcoal which retains all the impurities. The concentrations are then less than one ppm.

6.1.3 Ortho-para hydrogen conversion

At ambient temperature, the composition of normal hydrogen is 75% ortho-25% para. At 20 K, the composition of stable hydrogen is 99.8% para. Since the ortho-para conversion is exothermic, the pressure inside a normal liquid hydrogen tank would greatly increase. At 20 K the heat conversion which amounts to 703 J/g exceeds the latent heat of condensation. This amounts to only 444 J/g but this reaction heat increases when the temperature decreases. The ortho-para conversion is done at different temperature levels and we use a catalyst otherwise the reaction were too long. At liquid nitrogen temperature, the conversion is realized up to 45% para. Just before the last cooling, the conversion reaches 95% para (see Fig. 6.1).

6.1.4 Liquefaction

There are three successive stages in order to reach −253 °C: −40 °C (temperature of liquid ammonia), −196 °C (temperature of liquid nitrogen) and finally −253 °C (temperature of liquid hydrogen).

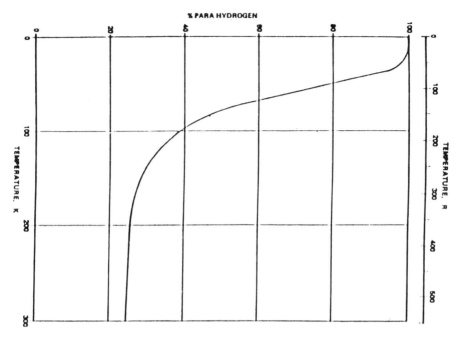

Fig. 6.1 Equilibrium ortho-para composition of hydrogen.

These temperatures are obtained in cold boxes which include the heat exchangers and the expansion turbines. The rotation speed of the gas bearings expansion turbines is higher than 200,000 rpm.

Liquid ammonia and liquid nitrogen are used as auxiliary refrigerating fluids which provide the necessary precooling at −40 °C and −196 °C. Ammonia and Nitrogen are liquefied in appropriate closed loop refrigerating cycles.

The feed hydrogen is cooled down and liquefied separately from the main hydrogen Claude cycle using different passages of the heat exchangers. By this way, the size of the adsorber and the catalysts can be kept at a minimum.

The last stage of the liquefaction in order to reach 20 K is achieved by a Joule–Thomson valve (see Fig. 6.2).

6.1.5 Efficiency

The specific liquefaction energy is nowadays about 0.950 kWh/l (or 13.4 kWh/kg) although the theoretical and reversible work is 4 kWh/kg with

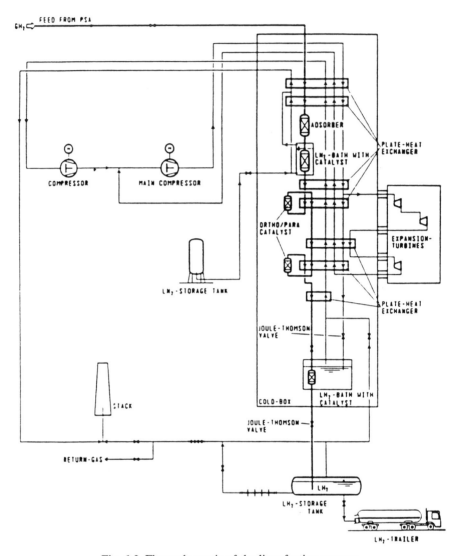

Fig. 6.2 Flow schematic of the liquefaction process.

the ortho-para conversion and 3.35 kWh/kg without it. The thermodynamic efficiency is estimated at 33% with respect to an ideal Carnot cycle with a four-stage ortho-para conversion. The losses are due to the compressors (29%), the N_2 refrigerator (25%), the turbines (13%) and the heat exchangers (13%).

6.2 PROCESSES UNDER R&TD

6.2.1 Slush hydrogen

In the different studies for the aerospace hypersonic plane (american project NASP for example), the envisaged fuel is not liquid hydrogen but slush hydrogen which is a mixture of solid and liquid hydrogen. This mixture would reduce the mass of the plane by about 20% because its density is more important. Its vaporization is longer and requires more energy so the plane can be cooled better. These elements are very interesting for an aerospace plane but not necessarily for an aircraft.

6.2.2 Other liquefaction principles

Other principles exist which liquefy gases at very low temperatures as in magnetic liquefaction. Any plant can use these processes which are used only in laboratories and do not seem to be interesting at an industrial scale and at temperatures higher than 20 K.

6.2.3 Efficiency

It is possible to study more elaborat cycles in order to increase the efficiency. For example: use the oxygen stream after a high pressure electrolysis to produce work and cooling; use other closed loop cycles for the precooling of hydrogen; adapt the purification to the hydrogen source. Many other ideas or patents could be explored.

6.2.4 Size

Hydrogen fuel requirements for an airport are estimated to 2250 t/d, i.e. 26.25 kg/s. Such a plant is 50 times more important than the biggest current plant (40 t/d). Early American studies had determined that a single liquefaction module having a capacity of at least 250 t/d is possible but larger sizes are not economically justifiable by reason of the compressors and exchangers efficiency.

In practice, the size of the current technology is limited to 40 t/d because of the H_2 reciprocating compressors the power of which reaches a maximum of 12 MW, the cold box which contains eight exchangers in parallel and the turbines (three in parallel for 10 t/d).

Though a study will have to develop such a plan with its own devices (compressors etc.), it seems to be more urgent to study the aircraft and the hydrogen management.

6.3 PRICE TRENDS

We have already seen that the global thermodynamic liquefaction efficiency is about 33% and that the ortho-para conversion increases the power consumption of 18.1% (reversible case). The major parameters are product para content, compressor efficency and recycle return pressure.

6.3.1 Conversion ortho-para

Figure 6.3 presents the theoretical works of the conversion in different cases. The conversion in one single stage at 20 K adds up 35.6% more work than when the conversion is carried out by continuous reversible means.

The most economical approach is to produce liquid hydrogen at a rela-

Process	Conversion Stage Temperatures, K	Work kj/g	kwh/lb
For 99.8% Para H_2			
1. Reversible	N.A.	14.28	1.799
2. Stagewise Plus Reversible	110	14.31	1.803
3. Stagewise Plus Reversible	80	14.42	1.817
4. Stagewise (5 Stages)	80, 65, 50, 35, 20.23	15.25	1.922
5. Stagewise (4 Stages)	80, 60, 40, 20.23	15.56	1.961
6. Stagewise (3 Stages)	80, 50, 20.23	15.74	1.983
7. Stagewise (2 Stages)	80, 20.23	17.53	2.209
8. Stagewise (1 Stage)	20.23	19.36	2.439
For Normal H_2 (25% Para)			
9. No Conversion		12.09	1.523
For 80 K Equilibrium H_2 (48.54% Para)			
10. Stagewise	80	12.55	1.581

Note: Feed is normal H_2 at 101.33 kPa (1 atm) and 300 K
Product is liquid H_2 at 101.33 kPa (1 atm) and 20.23 K
Heat rejection temperature = 300 K

Fig. 6.3 Theoretical work for liquefaction of hydrogen.

tively high concentration of ortho hydrogen using continuous conversion below liquid nitrogen temperature. A break-even time exists for which the energy cost for conversion equals the energy cost for the vaporized hydrogen (see Fig. 6.4). If the hydrogen is used within the break-even time limit (19 h for normal hydrogen and 36 h for 48.5% para-hydrogen), partial conversion is advantageous with respect to energy consumption. The knowledge of the meantime of storage could allow us to economize energy up to 15% over current liquefaction plants.

6.3.2 Process and devices

Parametric process variations were studied to determine the effect of process conditions changes. Results of variations in feed pressure, recycle return pressure, compressor and turbine efficiencies, minimum turbine refrigeration level and ambient temperature approach on the principal heat exchanger bank are given in Figs 6.5 to 6.9.

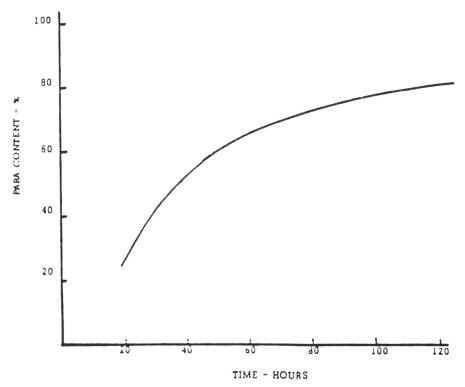

Fig. 6.4 Break-even time for partially converted liquid H_2.

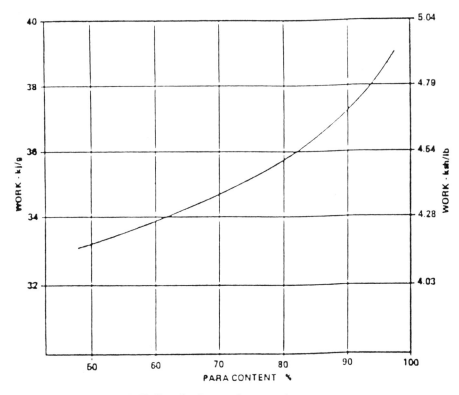

Fig. 6.5 H_2 liquefaction work vs. product para content.

Each percentage gain in efficiency of the compressors gives a 1.35% reduction in unit work but only 0.69% for the turbines. Unit work decreases with decreasing turbine refrigeration level because part of the lowest level refrigeration is diverted from Joule–Thomson to the most efficient work expansion method. The curve is terminated to 26 K to prevent liquefaction in the turbine exhaust stream.

In a recent study for a 40 t/d liquefaction plant of hydrogen produced from electrolysis, the total costs (warm and cold purification and liquefaction) are estimated to be:

- service capital: 57% (15 years payback, 8% interest: 11.7% annuity);
- electrical energy: 19% (hydroelectricity);
- operating costs (without energy):   24%;
- total energy consumption including auxiliaries: about 16 kWh/kg.

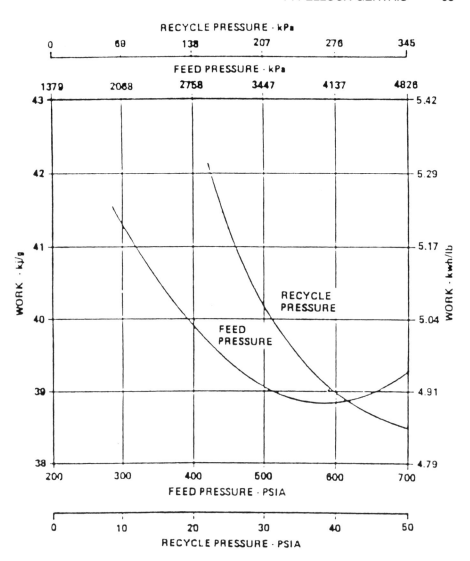

Fig. 6.6 H_2 liquefaction work vs. operating pressures.

6.3.3 2250 tonnes/day extrapolated plant

For a 2250 t/d plant, American studies had estimated the costs of the liquefaction to be:

- electric energy: 37.5%;
- cooler: 6%

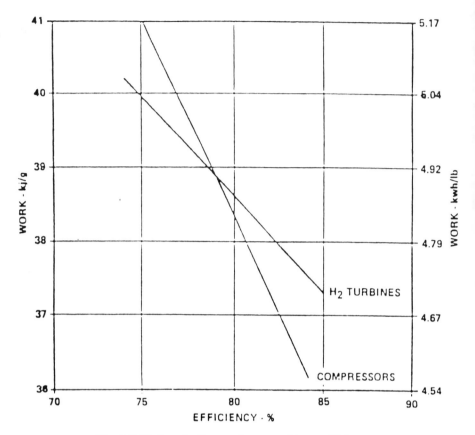

Fig. 6.7 H₂ liquefaction work vs. machinery efficiency.

- other operating costs: 12.5%
- service capital: 44% (annuity 11%);
- total energy consumption: about 12.50 kWh/kg;
- compressor efficiencies: 79%
- H_2 turbine efficiencies: 79%
- N_2 turbine efficiencies: 84%.

The cost of the electrical energy and the refrigerant does not depend on the dimensions of the plant.

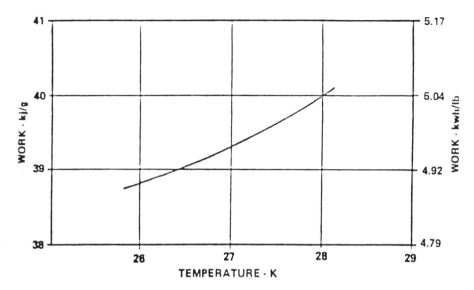

Fig. 6.8 H$_2$ liquefaction work vs. turbine refrigeration level.

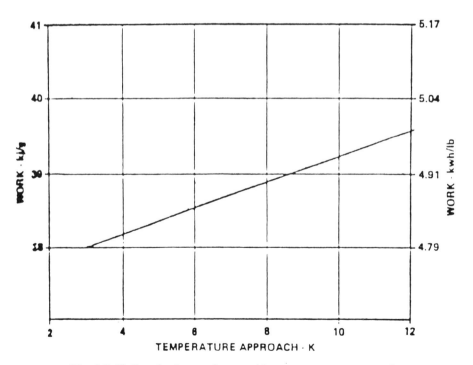

Fig. 6.9 H$_2$ liquefaction work vs. ambient temperature approach.

6.4 STORAGE AND DISTRIBUTION

6.4.1 Ground storage

In Kourou, the product is stored in five semi-mobile tanks of 320.000 litres each. These are tanks with a double shell whereby the annular space is under deep vacuum better than 10^{-6} bar and super insulation limits to the upmost heat losses.

L'Air Liquide at Waziers or Linde at Ingolstadt stores the product hydrogen in a vacuum insulated 270.000 litre tank which limits the losses to 0.3% per day. The two biggest European vessels of 600.000 litres each are used for the spatial engine and turbopump tests in France and Germany. In order to fill this tank from trucks, a 270.000 litre intermediate tank exists which is linked by a 300 m long and 150 mm diameter insulated duct.

In the plants of the L'Air Liquide's industrial customers, the liquid hydrogen vessels contain up to 60.000 litres with super insulation.

It is also possible to build tanks whose insulation were done by perlite and 10^{-2} bar vacuum. The losses per square metre are greater than with the super insulation. L'Air Liquide has defined a spherical tank of 14 m inside diameter for its plant in Becancourt, Canada.

6.4.2 Distribution

Nowadays liquid hydrogen is transported to the users by tank trucks which capacities are about 40.000 litres. These tanks are super insulated with a deep vacuum and anti-radiation barrier composed with aluminium sheets and glass felt. The distance of delivery can be thousands of kilometres. For example liquid hydrogen is transported by 32.000 litre containers to Kourou in the case of the Ariane 4 programme. This liquid hydrogen transport does not set any technical problem to the 'small' quantities.

To transfer the liquid hydrogen from a tank or a pipe to a vessel or a pipe, self pressurization or pumps are used. Since hydrogen is almost always used as a gas and not as a liquid, it is vapourized in a heat-exchanger just after the outlet of the tank so there is no liquid hydrogen pipe in the consumers installations. The running flowrates are about 10 to 1000 Nm^3/h in the ironworks or chemical plants.

We can find liquid hydrogen pipes only in production plants and in spatial installations. The length is limited to a few hundred metres in vacuum insulated pipes because the heat losses were too important in spite of the insulation.

For larger hydrogen consumption, there is in Belgium and the north of

France a gaseous hydrogen pipe network between Anvers (Belgium), Waziers and Isbergue (France). As for natural gas, it is possible to transport many thousands of cubic metres per hour of gaseous hydrogen over many large distances.

6.4.3 Ducts

The biggest diameter is now 254 mm (10 inches). This diameter does not seem to be a hindrance for the valves (butterfly valves) and the other components. The main problem of the big diameters is in the cooling which can cause the ducts to become unsymetrical and wrench the pipe. As soon as the diameter and the pressure are over 50 mm and 10 bar, compensators on the outer tube have been used in order to avoid the bottom effect.

6.4.4 R&TD

6.4.4.1 Insulation

To optimize the insulation of the vessels in the space projects (aerospace hypersonic plane, Ariane, Hermes etc.) and also in other applications, L'Air Liquide monitors and uses insulations other than vacuum super insulation or perlite such as honeycomb structure, foam, silica powder, molecular insulation etc.

With these insulations, we can adapt the insulation level to:

- the storage mean time;
- the mission profile;
- the mass;
- the outside temperature;
- the heat flux.

In the case of a plane, these parameters are important and justify a study which could also determine the need for an internal insulation, an external insulation or a mixed insulation. The insulation must also be adapted for:

- liquid hydrogen: temperature, compatibility etc.
- the design of the tank: shape, mechanical stress etc.

6.4.4.2 Fuel handling

The main difficulties when a cryogenic fluid is used or stored are:

- the vaporization of the liquid due to the heat losses;
- the over-pressures;
- the cooling of the pipe;
- the absence of oxygen and air;
- the low temperatures and the problem of cryogenic and hydrogen compatibilities;
- the heat exchangers;
- the safety;

and above all

- the management of the hydrogen in accordance with the mission profile.

For the space projects, L'Air Liquide has defined stringent rules to supply the engines, fill up the vessels and to store the cryogenic fuel. Adapted devices (pumps, valves, safety valves, heat exchangers etc.) have also been developed. We can also determine the best way to warm up liquid hydrogen (study of heat exchanger), if necessary, and to inject liquid or gaseous hydrogen in the engine.

In our case, it is necessary to study the hydrogen management according to the mission profile from the beginning of works so that we can define the safety rules, the handling procedures, the transfer, storage and engine feed installations, the insulation levels, the heat exchangers and the other devices (pumps, valves etc.).

6.4.4.3 Design of the tanks

Usually, the tanks are cylindrical or spherical. In the case of a plane, the vessel must be adapted to the design of the aircraft. It can even be structural and must then withstand the thermic and mechanical stresses. L'Air Liquide has studied such tanks for the space projects (Ariane 5, Hermes etc.). Important works must take note of the following:

- the shape of the vessel;
- the different materials: metals, composites etc.;

- the integration of the insulation;
- the mass;
- the thermic and mechanical calculations.

We can notice that L'Air Liquide studies 3600 m^3 tanks for the Euro-Québec project.

6.5 SAFETY RECORD

6.5.1 Introduction

We shall now discuss the implantation rules for LH_2 storage and distribution devices and the safety record.

When dealing with safety aspects in hydrogen it is important to bear in mind some of its relevant properties:

- flammability limits in air at 20 °C, 1 atm:

  —lower: 4%

  —upper: 75%;
- flame speed in case of deflagration: 3 m/s;
- flame speed in case of detonation: 2000 m/s;
- minimum autoignition temperature at 1 atm: 570 °C.

While hydrogen is a more hazardous gas than hydrocarbons, hydrogen vapour, thanks to its lightness and rapidity of diffusion, is more easily eliminated than petroleum gases. Moreover, in case of ignition, hydrogen fires, which are concentrated, rapid, and only slightly radiative, are less dangerous than hydrocarbon fires.

The best way to contain a hydrogen fire is to let it burn under control until the hydrogen flow can be stopped.

An accidental unstationary combustion is more likely to start than a slow deflagration. The passage from deflagration to detonation is promoted by turbulences which can be generated by obstacles like staircases, walls, etc. The turbulence enhances the heat transfer in the combustion which in turn increases the volumetric expansion.

The higher the speed of the reaction zone the higher the maximum overpressure in the combustion zone.

6.5.2 Results of a safety study for implantation of LH_2 storages

6.5.2.1 Approach used

The implantation rules for LH_2 storages and devices were determined using the following approach:

- Risk analysis for the devices and study of the environmental impact;
- Determination of the major events by fault tree analysis;
- Evaluation of the consequences of these events;
- Analysis of the selected scenarios, use of models (thermics, dispersion etc.).

6.5.2.2 Evaluation of the consequences of accidents

Following a loss of confinement of the liquid hydrogen, the effects expected upon the environment depend on the leakage ignition delay:

- Immediate ignition: only thermal effects have to be considered.
- Delayed ignition: an H_2/air mixture may form and be the source of mechanical effects.

In both cases, the bearable limit values differ between a human being or a building structure.

Thermal effects: The thermal effects due to an incident on the storage are linked to the development of a torch at the top of the vent stack following a blast of a rupture disc or a flame following a rupture of a liquid pipe.

- Structures: Two values were selected according to the exposure time:

    —Exposure time < 1 minute: 27 kW/m^2

    —Exposure time > 1 minute: 10 kW/m^2;

- Human being: The thermal flux of a fire may cause skin burns if the intensity of the radiation and the time of exposure are sufficiently important. The value of 5 kW/m^2 is considered as a dimensioning value for the human being protection.

Mechanical effects: The blast effect of an explosion of an H_2/air cloud depends on the intensity of the overpressure, the speed of increase of this pressure, and the time of the positive phase. The effects are more violent if the pressure increasing is quick (such as a shock wave). In the same way, damage is increased with the time it takes for the positive phase to reach at the end of an asymptotic value.

- Structure: The chosen value is between 50 to 70 mbar depending on the surrounding storage. Experimental results show that an overpressure of 50 mbar leads to the destruction of 75% of glass windows.
- Human being: The most fragile parts of a human body submitted to an explosion are the ear-drums and the lungs. The threshold value selected for the mechanical effects is 90 mbar. For the ear-drums, 90 mbar leads to 40% of rupture. For the lungs the burst begins at 700 mbar.

6.5.2.3 Determination of the implantation rules

To assess the consequences of the events considered as major ones a three dimensional calculation code is used. The dispersion of the cloud modelization due to a loss of confinement of hydrogen takes into account the buoyancy, the thermal and the turbulence effects.

The results lead to the implantation rules and safety distances to be observed around an LH_2 storage. Figures 6.10 and 6.11 give examples of the

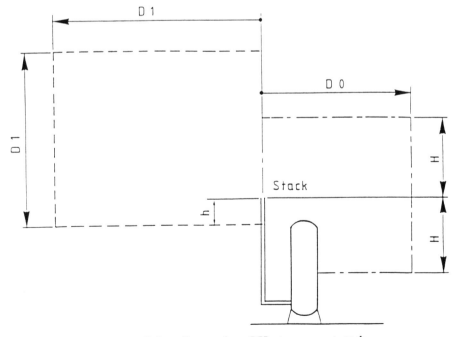

Fig. 6.10 Safety distance from LH_2 storage vent stack.

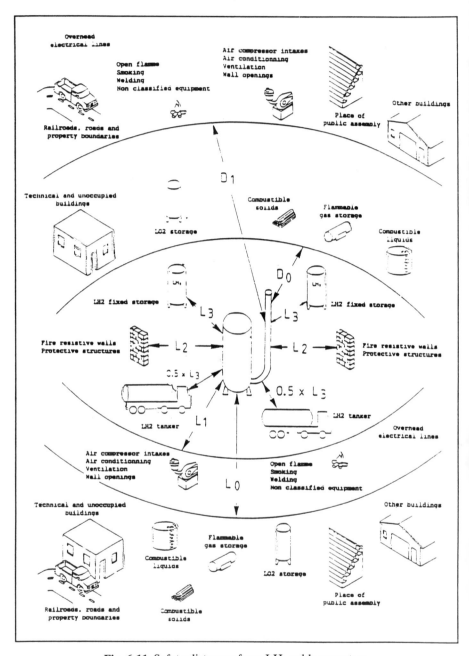

Fig. 6.11 Safety distances from LH_2 cold converters.

minimum safety distances from an LH_2 storage and its vent stack. The index '0' (L0, D0) relates to the flux criterion. The index '1' (L1, D1) relates to the flammability criterion and to the mechanical effects due to the explosion of the cloud.

These documents can measure precisely the distances according to the items in the surroundings of the zone.

For example the distance L0 is related to the event: 'Pressure build-up coil rupture'. If this event occurs, in case of an immediate ignition of the leak, the flame jet will have a diameter of 1 m and will be 13 m long. The thermal radiation hazards are limited to the immediate vicinity of the flame and the limit flux of 5 kW/m^2 is obtained at 18 metres from the tank.

In case of a delayed ignition of the leak, the cloud inflammation involves serious burns for the people present in the cloud or near it. A burn risk (second degree burns on unprotected skin) is possible in a 16 metres zone from the tank. Consequently we take: L0 = 18 metres.

Eventually, the implantation rules deal with equipment selection, based on the satisfaction following rules being satisfied:

- avoid confinement in the zone;
- safety zones markings;
- no naked flames, no sparks.

This also leads to equipment such as vent stacks, electrically safe devices, enclosure of the zone, water sprinklers etc.

6.6 REGULATIONS

To build liquid or gaseous installations, pressure vessels regulation transport regulations (A.D.R.) and classified installations regulations etc. must be respected. Other European countries have their own regulations or their own interpretations. They also use their own calculation codes which are not compatible with each other.

In fact, there is no specific regulation for liquid hydrogen which is considered only as a combustible gas or liquid. Nevertheless, its chemical or physical properties should require particular rules (for example for the materials).

Besides, current regulations are related to industrial utilization. They have not been written for ground or even air transport (liquid hydrogen handling in airports or for domestic cars or buses). Thus, an adequate regulation would have to be created.

6.7 CONCLUSIONS

The technology for cryofuel production exists at an industrial scale. The present process is well known even if some improvements are possible. The biggest current plant is, however, 50 times smaller than those required to supply an airport, so it is important to underline that compressors and other devices for such liquefaction plants do not yet exist.

As for the storage and the handling, some important applications exists which can serve as references for air/ground cryofuel.

6.8 RECOMMENDATIONS

Though compressors, heat-exchangers and big plants will have to be developed, it is still too soon to do that because the start-up of the plant will take place in 20 years. A study could however determine the more elaborate thermic cycles in order to increase the efficiency.

The most important work should concern the management of the liquid hydrogen according to the mission profile. So safety rules, handling procedures, transfer, storage and engine feed installations, insulation levels, heat-exchangers, pumps, flowmeters, level-sensors and others devices, and the injection into the engine can be defined.

Other studies should be related to the design of the aircraft or ground tanks (mechanical design, insulation etc.).

6.9 EXPERIENCE FROM LH$_2$ SPACE APPLICATION

6.9.1 Handling

6.9.1.1 Test facilities

Tests facilities devoted to the Vulcain rocket engine are equipped with specially designed high flow cryogenic ducts, which are used to feed the engine turbopump with liquid hydrogen with a flow up to 650 l/s, i.e. 45 kg/s. These pipes and tanks were developed by L'Air Liquide using the vacuum double wall technology with super insulation.

The characteristics of the pipes are:

- Length of the pipes: 39 m;
- Inside diameter: 254 mm (10 inches);
- Flowmeter cross-section: 304.8 mm (12 inches);

- Maximum service pressure: 15 bar abs;
- Overall pressure drop: 1 bar at a flow of 45 kg/s;
- Thermal leakage: 500 W;
- Tightness at 20 °C: 10^{-7} Ncm^3/s;
- Tightness at 196 °C: 10^{-6} Ncm^3/s;
- Shrinkage of the inner pipe at cooldown: 3 mm/m (comparator on the outertube);
- Valves: 10″ butterfly type valves with Cv 2500.

The pipes are equipped with many measurement points as temperature, pressure and flow.

Many requirements have to be respected for cooling down:

- minimum duct distorsions;
- minimum turbine flowmeter speed during the gaseous phase because these ball-bearings are designed for working in liquid hydrogen and not in gas which cannot cool them in good conditions;
- minimum cooling down consumption.

The major point during a test is to keep the inlet turbopump pressure at the specified value. It is controlled with a closed loop by a control valve. Due to the instability of the inlet turbopump pressure for the starting, it's the tank bottom pressure which is regulated.

Strict safety and handling rules exist to:

- fill the tanks;
- empty the vessels;
- cool the pipe;
- transfer liquid hydrogen;
- pump liquid hydrogen etc.

Most of the actions are automatic to suppress human error.

6.9.1.2 Ground installation in French Guiana

The flow and the diameter are less important than for the test facilities in Europe, only 4 and 6 inches for ELA 3. The length of the ELA 3 pipe is 350 m.

106 TECHNOLOGY FOR CRYOFUEL PRODUCTION

6.9.1.3 Ariane 4

The characteristics of the Ariane 4 H10+ liquid hydrogen tank are:

- capacities: 2036 kg;
- mass: 675 kg (with the LO_2 tank and the devices);
- flow: 2.62 kg/s;
- propulsive phase: 732.9 to 742.3 s.

Pressure regulation:

- ground: 2.7 to 3.35 bar abs with helium;
- flight: 2.8 to 3 bar abs with hydrogen.

The following parameters are controlled for the flight:

- hydrogen level;
- hydrogen temperature;
- wall temperature;
- pressure;
- vibrations.

The hydrogen is used for a rocket engine which consists of a turbopump providing the necessary energy to the propellants to feed the combustion chamber in which these propellants are burned. The turbopumps are driven by hot gas turbines. The hot gas is produced by a gas generator acting similarly to a small combustion chamber (see Figs 6.12 and 6.13).

6.9.2 Safety record

From the beginning of the Ariane project, there were no problems with safety when using liquid or gaseous hydrogen and especially:

- no hydrogen leak;
- no pressure rise due to insulation problem or regulation problem;
- no explosion;
- no structure ruin;
- no problem due to the corrosion under stress in presence of hydrogen.

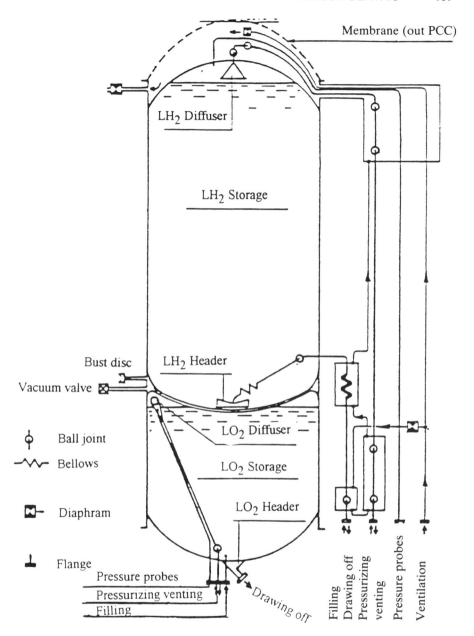

Fig. 6.12 Ariane 4 H10+ storage layout.

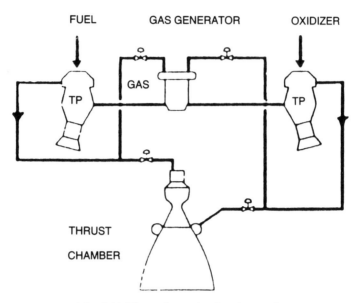

Fig. 6.13 Flow schematic of rocket engine.

In fact, the handling of liquid or gaseous hydrogen has not resulted in any incidents to date.

6.9.3 Components applicable in air/ground transportation

In the space applications, many sensors (pressure gauge, temperature probes, level indicators, flowmeters etc.) and many components (valves, pressure safety valves, pumps, turbopumps, pressure regulator etc.) are used. Generally, this equipment has been developed specially for Ariane because it has to withstand strict specifications (precision, vibration level, acceleration, environment, reliability etc.). Finally, most of the equipment cannot be used without adjustment. However, we can note that this type of component has been proved and that the methods of measurement and the type of component such as the turbopump can be used for air/ground transportation.

6.9.4 Conclusions, recommendations

Spatial experiences will be very useful for the development of liquid hydrogen fuel and solve important problems. But many important differences exist because liquid hydrogen has yet to be used in public locations (airports, petrol stations, cars etc.) by non-specialists.

It is thus important to continue safety studies and procedures of all aspects of liquid hydrogen use to ensure its reliability in the public domain.

7
Experience from LH$_2$ Space Application

M. Müller, F. Grafwallner, P. Luger
DASA Raumtransportsysteme und Antriebe, Ottobrunn, Germany

7.1 INTRODUCTION

Hydrogen as fuel and oxygen as oxidizer has become a standard propellant combination for modern space launchers. Typical examples are the Apollo Moon Flight Launcher (2nd and 3rd stage), the Centaur rocket (upper stage, United States), the US Space Shuttle, the Russian Space Shuttle carrier 'Energia', the Japanese H-2 rocket, the European Launcher Ariane III/IV (3rd stage) and the future European Launcher Ariane V. Also the future hypersonic airbreathing space planes will be exclusively operated with hydrogen.

The only reason for the selection of hydrogen has been the attractive high specific thrust (thrust force per fuel mass consumption) for both rocket and space plane propulsion, because the propellant fraction of spacecraft is unusually high (about 90% for a typical rocket launcher). Environmental aspects have not been considered in the past, but the favourable emissions from hydrogen combustion will be very helpful for the implementation of future projects.

7.2 ROCKET ENGINE TESTING

Spacecraft need pressureless and liquid hydrogen because of weight and size restrictions for the propellant tanks. Consequently the development and testing of the engines has to be performed with liquid hydrogen too, with respect to thermal and density conditions. A typical rocket engine flow schematic comprises the following main components (Fig. 7.1): the turbine

112 EXPERIENCE FROM LH$_2$ SPACE APPLICATION

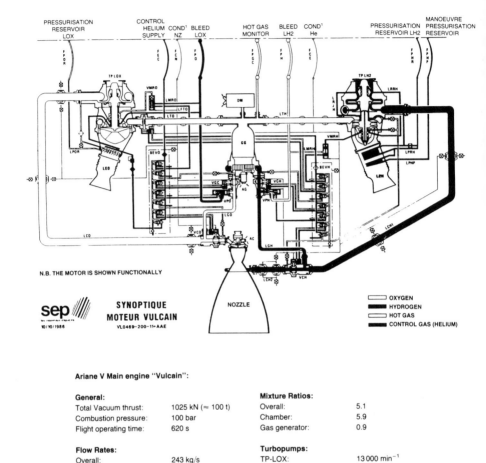

Fig. 7.1 Typical rocket engine flow schematic and performance data.

Ariane V Main engine "Vulcain":

General:		Mixture Ratios:	
Total Vacuum thrust:	1025 kN ($\approx$ 100 t)	Overall:	5.1
Combustion pressure:	100 bar	Chamber:	5.9
Flight operating time:	620 s	Gas generator:	0.9
Flow Rates:		**Turbopumps:**	
Overall:	243 kg/s	TP-LOX:	13 000 min^{-1}
Chamber:	231 kg/s		2.9 MW
Nozzle cooling:	1.8 kg/s	TP-LH2:	34 200 min^{-1}
Gas generator:	8.1 kg/s		11.3 MW
Auxiliaries:	1.5 kg/s		

driven centrifugal pumps for liquid hydrogen (LH$_2$) and liquid oxygen (LOX), the gas generator for the turbine supply, several valves and the thrust chamber consisting of the injector, the combustion chamber and the expansion nozzle. These components have to be developed to a high degree of reliability, before being integrated into the complete engine for further system testing. Figure 7.2 shows the basic flow diagram of a test facility for the subsystem 'thrust chamber' (injector + combustion chamber + expansion nozzle). The real flow schematic is much more complex with provisions for precooling, purging, ventlines, secondary systems and so on.

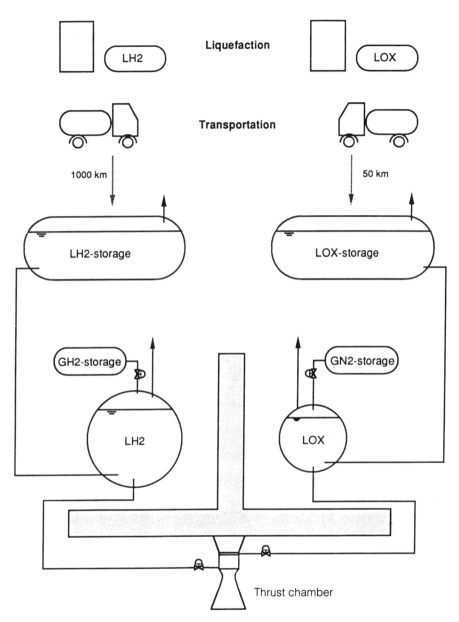

Fig. 7.2 Basic flow diagram of a rocket thrust chamber test facility.

The basic flow diagram (Fig. 7.2) shows symmetric installations for LH_2 and LOX. Both propellants are brought to the facility by road trailers: LOX from different production places in Germany and LH_2 from liquefaction plants in Germany and in France over distances of up to 1000 km. These transportations are performed according to standard rules, without police escort or other special procedures.

Unloading from the trailer to the storage tank is a routine procedure with purging, evacuating and precooling of the coupling before the flow is started. LOX-trailers are generally equipped with centrifugal pumps for the flow transfer into the storage tank. Modern LH_2-trailers have centrifugal pumps and additionally a conventional overpressure transfer system. There is some evidence, that these pumps still need additional development effort.

The unloading procedure, in principal, has been unchanged since the beginning of cryogenic handling. So there is a considerable potential for saving time and manpower by using advanced couplings and by the implementation of a computer equipped control system.

Trailers, storage tanks and transfer pipes, including necessary flexible pipes are vacuum insulated in order to avoid boil-off losses and to minimize the tank pressure increase. During the 1000 km transfer of the LH_2 no pressure release is necessary. The loss rate for typical storage tanks is 0.1–0.5% of the maximum tank capacity within 24 hours, the smaller value being valid for larger tanks, the higher one for smaller tanks.

The injection of the propellants into the thrust chamber (combustion pressure 100 bar) requires corresponding high propellant pressures. Usually the engine pumps are not very available, because they are being developed separately. The required pressure is therefore generated by pressurization of the so called 'run tanks' by means of high pressure gas.

The high pressure gaseous hydrogen from a pressure bottle (max. 800 bar) passes a pressure regulator and is fed into the LH_2 run tank (max. 400 bar). A special inducer avoids turbulence and mixing with the liquid. The LOX run tank is pressurized correspondingly, but with nitrogen gas instead of gaseous oxygen, in order to avoid the well-known problems (reaction with organic materials, metal particles and even small metal elements).

Through the high pressure feedlines – equipped with control valves – the propellants flow to the thrust chamber, where they are ignited by means of a pyrotechnical or electrical igniter. These feedlines do not need any heat insulation, because the two phase flow condition is suppressed by the high pressure level and because the temperature increase between tank and thrust chamber is small due to the high propellant velocity.

Because of the very expensive high pressure tanks and bottles, the test duration is limited to 20 seconds for the 100 tonnes to thrust chamber. Modern measurement and control equipment however, allows 10 different steps to run for thrust or mixture ratio within this short time. Long duration testing is performed later on a system level within the integrated engine.

The main data of the Vulcain engine are listed in Fig. 7.1. Other hydrogen operated DASA facilities are:

- gas generator test facility, similar to the thrust chamber facility, but smaller scale (450 MW bulk thermal power),
- LOX pumps test facility with turbine drive (3 MW mechanical power, 450 MW bulk thermal power),
- test facility for aircraft combustion chambers with LH_2-supply,
- test facility for hypersonic H_2-ramjet engines with thermal simulation up to $Ma = 7$,
- several smaller facilities for cryogenic component tests (tanks, valves, etc.).

7.3 TEST FACILITIES

The introduction of hydrogen into the Ottobrunn Test Field was started with very small rocket combustion chambers for a technological programme in 1962. Only gaseous hydrogen was used, with ambient temperature and a flowrate of max. 20 g/sec. The low combustion pressure of 15 bar allowed hydrogen supply from standard pressure bottles. Cooling of the combustor wall was performed with water. The test conditions were thus kept at a low level risk, and all test operations were run without any incident.

The next step included the supply with liquid hydrogen and an increase of flowrate up to 300 g/sec. A new test facility with two test positions and a liquefaction plant had to be designed and realized. In some places liquefaction of hydrogen was necessary, because no authority regulations for the transportation of LH_2 existed at that time.

Safety-related design aspects for the new buildings were:

- separation of test facility and liquefaction plant, but integrated operation (only one tank for liquefaction and test),
- open door and roof for ventilation of the test boxes during the hot firing of the combustors,
- permanent ventilation (20 times per hour) for the liquefaction plant,
- four separate boxes for the different subsystems of the liquefaction plant, as well as light roofs and large doors for relief in case of explosions,

116   EXPERIENCE FROM LH₂ SPACE APPLICATION

- automatic gas detectors on many places, including ventilation in the exits,
- any electric installation in explosion proof quality,
- test personnel protected from the test cell by a concrete wall and rigid observation windows.

The technical equipment was designed according to existing general rules and codes for pressure vessels and pipes. Insulated pipe connections were of the proven Johnston type, valves were developed together with a small, but flexible company.

After correction of small technical problems the new facility worked well for about ten years. After four years of operation the liquefaction could be stopped, when transportation of LH₂ on the road was licensed and a new Linde-liquefier was made available at a distance of 30 km. It is interesting to note, that the first transportation vehicles at that time had a capacity of only 1000 litres.

Meanwhile two significant larger facilities have been designed by the same engineering and test team and are intensively in use:

- 300 kN (30 tonnes) thrust chamber test facility at Ottobrunn,
- 1500 kN (150 tonnes) thrust chamber test facility at Lampoldshausen.

Figure 7.3 shows the evolution of pressure rates over 30 years of permanent

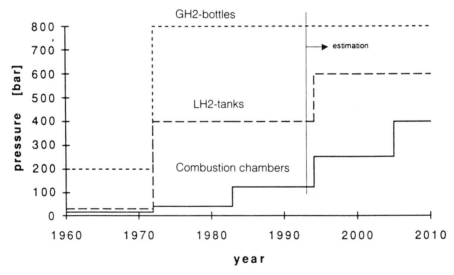

Fig. 7.3 Pressure evolution of DASA hydrogen test equipment.

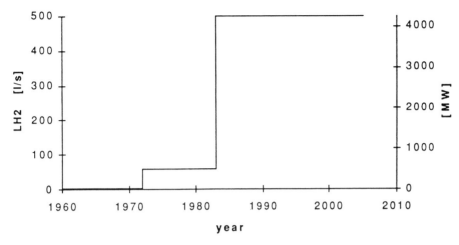

Fig. 7.4 Evolution of flow rates and global thermal performance.

practical work with both gaseous and liquid hydrogen. The next significant step in both combustion pressure and tank pressure has already been started. A small-scale facility with a combustion pressure capability of 400 bar is just being designed (by DASA and will be available in 1995. Figure 7.4 shows the evolution of LH_2-flow rates up to the present stage of 500 l/sec. About 75% of the corresponding energy content is realized within the combustion chamber with a hydrogen-rich mixture ratio of approximately 6 (oxygen/hydrogen masses), 25% of the hydrogen reacts with the ambient air along the exhaust jet. The right scale of Fig. 7.3 shows the considerable amount of total energy available in these facilities.

7.4 EXPERIENCE

Only general authority rules were available in the early days of our hydrogen facility activities. Some hydrogen specific data were taken from the more advanced experience in the USA, where hydrogen already had its high priority for future space projects. Today the corresponding authority laws and rules are much more detailed and specific. But it is still very helpful for the procedure of a new facility licence, if comprehensive and successful experience is available.

For safe design and operation of rocket engine facilities the following general main rules are essential:

- only persons who are trained and experienced with hydrogen and who are familiar with the facility should be engaged,

- sufficient number of data records and their critical review will help to detect anomalies or degradation of both engine and test facility,
- open air facilities avoid critical leakage concentrations and will limit the consequence of an ignition to a fire only instead of explosion,
- components and system design must have a high standard of engineering skill and care,
- close observance of maintenance periods and procedures.

During the 30 years of continuous DASA activities with hydrogen only very few small incidents have occurred, all of them without any injuries to people. It should be noticed that rocket engine facilities, in comparison with a possible hydrogen application for automobiles or aircraft, have to cover special additional risks like open fire (exhaust jet), very high pressures and unusually high flowrates.

Furthermore, testing in general means very frequent modifications of the experimental equipment because of changing test requirements. After our experience, these changes have always been the major source of possible incidents. Although new set-ups are always checked very carefully after their installation, there is always a risk when starting a new operation. The following incidents occurred on DASA H_2 rocket engine facilities (and are typically always linked with the first operation of new facility):

- During a combustor test with necessary rapid cooldown of a high pressure piping section one of the pipe connections leaked considerably. A combustible mixture of hydrogen and air was ignited and deflagrated with damage to a number of cables. A later investigation showed, that this standard pipe connection was leak tight in warm and cold conditions, but leaked during the cooldown process because of the non isothermal condition. Improved design of the connection was possible.

- During a small modification work (soldering) on an engine installed on the facility, a deflagration occurred close to working people, but without injuring them. Despite thorough purge procedures, which are mandatory for such repair work, a small hydrogen leak led to a combustible mixture. Improved purging procedures and the application of additional temporary gas detectors were the consequence.

- During the start-up phase of a thrust chamber with the sequence: hydrogen → igniter → oxygen, the ignition detection signal was positive and allowed the oxygen valve to open. This detection signal however was wrong, and the well-mixed (from the injector) propellant jet was ignited by a pilot flame at the end of the supersonic diffusor, necessary for altitude simulation. The engine and its surrounding vacuum capsule were completely destroyed by overpressure. Improved electronics and redundancy for the ignition detection system were the consequence. It has to be noted however, that this incident was not caused by the hydrogen itself, because a gasoline-air mixture would have led to the same result. But this event demonstrates the potential of available sensors, electronics and computer systems for safety applications.

From incidents in other places and countries we learnt about the following very important safety related phenomena:

- *Material embrittlement:* It was observed, that pressure cylinders for GH_2 developed small cracks due to embrittlement of the material at ambient temperature. The result of comprehensive research was, that only the very clean GH_2 from evaporated liquid is able to penetrate into the grain structure of the proven steel with subsequent embrittlement. The problem can be overcome by the addition of impurities to the GH_2, reduction of material stress, change of material or application of a liner made of less sensible material (DASA 800 bar GH_2 bottles).

- *Oxygen impurities:* Small contaminations of air in liquid hydrogen freeze into solid particles. Impurities of the introduced high pressure gas can thus accumulate in the pressure tanks in the form of solid nitrogen and solid oxygen. A mixture of solid oxygen and LH_2 however is explosive and needs only little energy for ignition, for example a small pressure shock or even the flow through a valve. Bursting pipes and detonation-like pressure and increase of high pressure LH_2 tanks have been observed. It is common practice at DASA, to warm up all LH_2 tanks periodically once or twice a year above the oxygen liquefaction temperature in order to vaporize the impurities with subsequent purging. Additionally, the gas used for pressurization of the tanks is under permanent online control (measurement of oxygen content).

The Tupolev Company reports, that during the development of their LH_2 experimental aircraft, a system has been developed, which retains any solid particles during the refuelling procedure of the aircraft tanks.

7.5 COMPONENTS APPLICABLE IN AIR/GROUND TRANSPORTATION

In principle the LH_2 supply system of a spacecraft is very similar to the corresponding supply system of an LH_2 aircraft. In both cases the LH_2 has to be stored in large tanks and to be transferred with a considerable high flowrate to the engines and to be pumped there to a considerable high pressure level. Reliability and lightweight are obligatory requirements for both applications. In the case of LH_2 automobiles the tanks and the flowrates are significantly smaller, but the desired pressure rise for injection is still high. Weight considerations have secondary priority. For these reasons the LH_2 components for automobiles have to be very different from spacecraft or aircraft applications. A more precise comparison of spacecraft and aircraft cryo-components (Table 7.1) shows some remarkable differences of their requirements with corresponding consequences for different design features.

Despite these differences however, there is the very strong commonality due to the fact that LH_2 components – when compared with conventional kerosene-components – need a special and very different design due to the following reasons:

- metal materials must be resistant against embrittlement,
- hard seals are not applicable because of high sealing forces,
- conventional soft sealing materials cannot be applied, because the get hard at low temperature,
- shrinkage of materials is considerable and has to be respected in special design,
- with sliding material pairs, the frictional behaviour is completely different at low temperature,
- heat insulation provisions have to designed and may affect the accessibility for maintenance and repair,
- hydrogen is a very poor lubricant in comparison with kerosene.

This demonstrates, that LH_2 components design for aircraft and other applications has to be based on existing experience with both space-borne and terrestrial hardware.

Table 7.1 Comparison of spacecraft and aircraft cryo-components

	Spacecraft	Aircraft
Lifetime	3 h	1,000,000 h
Mission time	0.15 h	40,000 h
Maintainability	moderate	moderate
Price economy	moderate	high

7.6 RECOMMENDATIONS

General: Although handling hydrogen is almost routine on our test facilities, there is still quite a list of wishes that remain. As the task is essentially always the same in most applications of hydrogen, we think that these proposals are of general interest for the whole field of hydrogen utilization. From our point of view the most important would be:

- automatic unloading of LH_2 from transportation vehicles instead of hand operation; for LH_2 automobiles the first step towards such a device has been reached;
- transfer of LH_2 between containers or from a tank to consumers should generally be performed by pumps; the conventional method by tank pressurization takes much time, creates vapour loss and allows only limited fluid velocities;
- a simple but efficient method to avoid or separate and to detect critical impurities within the LH_2 should be developed in order to avoid operational and safety problems;
- a small and economical device for re-liquefaction of evaporated hydrogen or even avoidance of evaporation would be helpful for a number of LH_2 storage applications.

Aircraft application (cryoplane): The DASA-contribution to the cryoplane feasibility study was concentrated on the onboard fuel system and therefore could make favourable use of the existing theoretical and practical knowledge, because the fuel supply system for a rocket engine is in principle very similar to that of an aircraft engine. One of the primary results of the study was, that no basic problems were detected, which could result in big difficulties or could even make it unfeasible. It is necessary however, to modify the existing proven components considerably for the application in an aircraft, because:

- the required lifetime is larger by orders of magnitude,
- the production cost has to be reduced considerably,
- maintenance and repair require different design principles,
- procedures have to be simplified and to be automatically executed.

Therefore, the following activities are proposed as a necessary step before starting the aircraft development program:

1. development of progressive tankwall structure and insulation methods;
2. improved technology for long-life pumps with optimized suction behaviour;
3. improved technology for long-life and absolutely leak-tight valves;
4. design and pre-development of an aircraft type gas compression device, which will allow the use of low pressure 'waste hydrogen' for the engines and the APU (auxiliary power unit);
5. design and pre-development of a coupling with easy cooldown characteristics for high flowrates;
6. definition of a practical system to prevent impurities in hydrogen and the detection of them (oxygen, nitrogen);
7. mathematical modelling of LH_2 systems, because of the more complex fluid compared to kerosene and gasoline (boiling liquid, low temperature, in general close to the critical point);
8. safety studies, theoretical and practical, in order to convince the public that hydrogen can be used without any increase of risk.

It is obvious, that these technological objectives are very much the same for conventional space propulsion, future space transportation and hydrogen aircraft. Additionally it can be assumed, that most of the listed items are also valid for any ground traffic application too. Therefore, a high degree of dual or even multiple use can be expected from a technological programme of the proposed type.

8
Experience from the TU-155 Experimental Aircraft

A. Shengardt, V. Sulimenkov, V. Malychev, V. Borisov
Tupolev Design Bureau, Moscow, Russia

8.1 INTRODUCTION

Providing sufficient energy will become a problem in the near future due to depletion of natural, fossil resources, at least in particular regions of the world. Moreover, the ecological situation is deteriorating because, as a result of human activities, vast amounts of pollutants have been released into the atmosphere since the beginning of the industrialization. Particularly, the use of hydrocarbon energy sources increases the atmospheric concentration of the greenhouse gas carbon dioxide (CO_2).

These problems cause more and more countries, especially the highly developed from the technological point of view, to start research and development activities directed to hydrogen technology. As an alternative to fossil resources, hydrogen is considered to be the ecologically pure universal energy carrier for the future. The transition to hydrogen in transportation, industry and other fields of energy utilization is one way to the radical solution of the atmospheric pollution problems, caused by carbon dioxides, nitrogen oxides, sulphur oxides etc. As hydrogen will be produced from water by thermal decomposition or electrolysis and the combustion product, again, is water, a hydrogen system can be considered as a closed cycle without any changes to the planet's water balance.

Taking into account that it will take decades of components development, infrastructure build-up etc., the German-Russian Cryoplane project, aiming to utilize liquid hydrogen (LH_2) as aviation fuel, is a timely step towards a hydrogen system. For some regions of the world with highly developed technology, it may be logical to change now from traditional transportation fuels to hydrogen. In other regions with large deposits of natural gas, the use of liquefied natural gas (LNG), which is similar to hydrogen as a cryogenic fuel could be a meaningful intermediate step. It also offers some advantages towards environmental protection.

8.2 THE TUPOLEV COMPANY'S PHILOSOPHY USING HYDROGEN IN TRANSPORTATION

The global energy demand, which at present is covered mostly by limited fossil resources, is rapidly increasing. According to the current state of knowledge, oil reserves will be used up within some 40 years on a global scale, in particular regions like the CIS this could be as early as at the beginning of the coming century.

The future energy demands may be mostly satisfied by nuclear fission energy, thermo-nuclear energy, which could be realized in the coming decades, and, in the long term, renewable energies, with hydrogen as the most suitable energy carrier.

Hydrogen, today, is used in many branches of industry, e.g. oil refining, ammonia and methanol production, metallurgy, nuclear and space engineering (Shuttle, Buran, Ariane etc.). The recent consideration of hydrogen as an energy carrier/fuel for transportation is driven mainly by its energy content (in liquefied condition nearly three times as high as with hydrocarbon fuels) and its environmental compatibility, as hydrogen avoids all pollutants related to hydrocarbon fuels like carbon oxides, sulphur and sulphur oxides, cancerous substances etc. Particularly, the CO_2 accumulation in the atmosphere is the major contribution to the enhanced greenhouse effect, which, if not stabilized in the near future, may lead to catastrophic consequences to the planet. The only pollutants emitted when hydrogen is used as a fuel in thermal power plants are nitrogen oxides, formed at combustion temperatures in excess of 1700 °C.

In the long term, hydrogen will be produced by electrolysis of water, powered by the mentioned nuclear or renewable primary energies. The oxygen, which is produced simultaneously, could e.g. be used for ecologically compatible waste removal. In the short term, hydrogen can be obtained by CO_2-free decomposition of natural gas and, thus, will be able to begin substituting non-renewable energy carriers in the near future. As an energy carrier, hydrogen offers the opportunity to transfer energy over long distances with high efficiency, even superior to electric power transmission, for example.

In automotive transportation the use of hydrogen will essentially improve the planet's ecology, especially in cities/conurbations and areas near highways with heavy traffic, where automotive transportation is the main contributor to environmental pollution. Preferably, hydrogen should be used in the liquefied, cryogenic condition, because:

- both the weight and volume-related energy contents are superior to hydride and compressed gas storage;
- the huge cooling potential may be used to substantially improve the technical performance of combustion engines.

The main problems to be solved for the wide use of LH_2 in automotive transportation are:

- the design of relatively cheap cryogenic fuel tanks with effective heat insulation, to store LH_2 in vehicles for a reasonable time without evaporation;
- the modification/redesign of combustion engines in order to increase the efficiency and to minimize emissions of nitrogen oxides.

At present, experimental passenger cars with cryogenic tanks are being operated in Germany and Japan, the storage time of LH_2 is up to several days. These tanks, having a screened-vacuum super insulation, are very expensive, can be easily damaged and have poor maintainability characteristics. If the on-board LH_2 storage time may be reduced, which is possible for some types of lorries and buses, the cryogenic tanks may be substantially cheaper by using less effective but more conventional kinds of insulation, e.g. polyurethane foam.

Using LH_2 could also be an alternative for railway transportation, particularly in countries like the CIS, where vast distances are to be covered which excludes electrification of the railroad network. LH_2 may be stored in separate cisterns already available in Russia which could be mounted on the locomotive.

The use of LH_2 in water transportation, besides the engine modification, will require safe accommodation of the cryogenic tanks with good ventilation of compartments containing cryogenic equipment.

Air transportation contributes only 2 to 3% to global energy consumption. However, air traffic pollutions are emitted within extremely sensitive layers of the atmosphere, which requires the transition to alternative, clean fuels in air transportation. Moreover, in the vicinity of major airfields air traffic contributions to local pollution are in the same order as ground traffic emissions.

The possibility of using LH_2 in aviation was proved convincingly by the TU-155 experimental aircraft, developed by the Tupolev Design Bureau. One of its three powerplants was operated on hydrogen only. The amount of

LH_2 stored on board allowed engine operation up to 1 hr. The development and testing of this aircraft found ways for engine and cryogenic system design to meet the aviation requirements, and to work out operational procedures that are similar to the analogue procedures of conventional kerosene aircraft.

The further joint work of Tupolev and Deutsche Aerospace Airbus on the Cryoplane project proved the feasibility of hydrogen cryosystems in passenger aircraft. It opens the way to the development of aircraft structures, cryosystem components such as tanks, pipes, heat insulation, pumps and safety systems.

In the public opinion, doubts are sometimes expressed regarding the safety aspects of hydrogen use. This can be explained by school knowledge of oxyhydrogen gas explosions and by the memories of the accident of the dirigible 'Hindenburg', filled with hydrogen gas. On the other hand, a large number of barrage balloons filled with hydrogen were widely used during the Second World War without any problems, explosions or fires. At present, hydrogen and its properties are known well enough and ways and means have been developed for hydrogen use with the same safety levels as traditional fuels of organic origin. In case of a hydrogen fire, structural damage and human injuries would be less because of the rapid evaporation and volatilization of hydrogen, the high burning velocity and the low radiation of the flame in comparison with hydrocarbon fuels.

It can be concluded that the technological level of the industrialized countries today could mean the beginning of a transportation system using LH_2 as a fuel.

A crucial problem in this process is the development of an infrastructure for production, distribution and storage of LH_2 and for refuelling vehicles. As this infrastructure will substitute for the traditional infrastructure in the long term, it has to be ecologically clean, efficient, foolproof and comparable to today's infrastructure regarding operating costs.

The process of finding hydrogen consumers and means for its production, distribution and supply must be coordinated and developed integrally.

8.3 THE TU-155 EXPERIMENTAL AIRCRAFT

8.3.1 General description

Driven by the energy crisis in the 1970s, a feasibility study on alternative cryogenic fuels for transportation, including aviation, was carried out in Russia. As a result of this study the design of an experimental aircraft operating with these alternative fuels was started at the Tupolev Design Bureau.

The TU-154 passenger transport was selected as the baseline aircraft. It is equipped with three turbofan engines and has demonstrated a high level of reliability during long-term intensive operation.

The main aims of the experimental aircraft were to study the possibility of aircraft operation using cryogenic fuels and to determine ways of developing aircraft cryogenic fuel systems and components. On the TU-155 experimental aircraft (Fig. 8.1), the right-hand powerplant, including the engine and its fuel and control systems, was modified for operation on liquid hydrogen or liquid natural gas alternatively. A cryogenic fuel tank was installed in the rear section of the cabin. The other two powerplants remained unchanged, operating on kerosene.

During the process of cryogenic powerplant development the experience gained in cryogenic ground facilities and rocket powerplant development was incorporated. Special attention was paid to provide maximum explosion and fire safety. However, it has proved impossible to design the aircraft

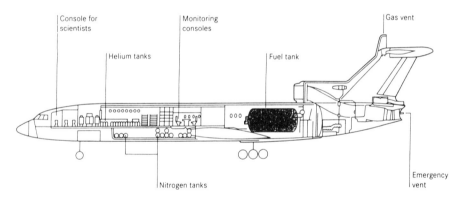

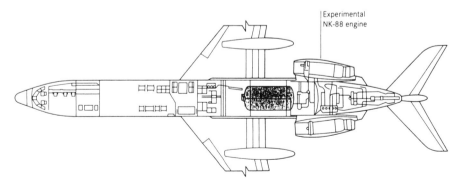

Fig. 8.1 TU-155 experimental aircraft.

cryogenic powerplant using only well-known principles and experience from rockets and ground facilities. Aircraft operating conditions are essentially different, regarding:

- operational life which is in the order of 10,000s of hours instead of minutes;
- engine start-up which is accomplished at small fuel flows;
- weight and geometry restrictions of system components which are insignificant for ground facilities.

The following major components had to be developed for the cryogenic fuel system:

- the fuel tank with a separated supply tank inside (capacity 1.5 m^3), where three centrifugal pumps and anti-overload devices are installed. It provides reliable engine fuel supply over the whole range of operating regimes;
- devices providing cryogenic fuel supply in boiling conditions;
- engine fuel supply pump system with electrically driven centrifugal and jet pumps;
- light structure of cryogenic valves and joints;
- a fuel circulation system through the turbopump assembly for cooling before starting the engine;
- a fuel heat-exchanger for evaporation and heating of the fuel prior to injection in the combustion chamber.

8.3.2 Components

The cryogenic fuel tank with a total volume of 18 m^3, installed inside the fuselage tail section, was manufactured according to the technological methods for transport cisterns design with a screened vacuum heat insulation. The inner vessel is made of austenitic stainless steel with a relative mass of 180 kg/m^2. The absolute operating pressure in the tank is maintained within the limits of 1.3–1.9 kg/m^2.

The fuel lines are made of stainless steel and also have screened vacuum heat insulation. The weight per metre amounts to 10–12 kg. The fuel lines are connected to other system components by welding.

The fuel tanks and pipes are a 'tube-into-tube' system. Hydrogen leakages into the vacuum chamber are controlled by temperature indicators between inner and outer tube. A safety device protects the system against a pressure rise above 3.0 kg/m^2.

The fuel system is remotely controlled and activated by compressed helium with a pressure of 80–100 kg/m^2. Two remotely controlled shut-off valves are installed on each fuel pipe. One valve is installed inside the fuel tank providing the possibility to drain the pipe fully from cryogenic fuel by helium bleeding. The second valve is installed at the end of the pipe. If the fuel pipe is not in operation, the volume between the valves is filled with helium. The inner vessel of the cryogenic fuel tank has two drainage systems – an operational and an emergency system – bled by helium. The vacuum chamber has a protection membrane connected to the drainage system.

In all compartments equipped with cryogenic system components, sensors informing the crew about gas concentration are installed. These compartments may be filled by nitrogen or ventilated by air.

In the engine, the cryogenic fuel system exists in parallel with the kerosene one. During engine operation on cryogenic fuel, the kerosene is recirculated to the tanks by the conventional kerosene system. The cryogenic fuel is supplied to the combusters by a turbopump assembly, driven by compressor-bled air. The fuel pipes within the engine nacelle are polyurethane-foam insulated.

8.3.3 Flight test and ground facilities

The experimental aircraft was successfully tested, operating on both LH$_2$ and LNG. Six flights on LH$_2$ were conducted, during which the cryogenic engine was operated under all aircraft manoeuvers and thrust regimes from take-off power to idle, including in-flight shut-down and restart. Maximum engine operating time with cryogenic fuel is about one hour per flight.

Operating on LNG, the TU-155 made several demonstration flights to the international airports of Moscow, Bratislava, Nice, Berlin and Hannover. Up to now, the aircraft has accumulated some 100 flight hours. The TU-155 tests showed the real possibility of developing and operating aircraft on LH$_2$ and LNG. Work on the TU-155 project cleared the way to design cryogenic powerplants and systems with weights comparable to the kerosene system weights. The cryogenic aircraft cost efficiency will be determined by the cryogenic fuel cost.

The activities at the Tupolev Design Bureau provided not only the TU-155 flying laboratory development but also test stands and test procedures. Essential topics comprise:

- test stands

    —for low temperature mechanical tests of material samples;

- for tests of fire extinguishing means in the cryo-aircraft;
- for strength tests of cryogenic model tanks;
- for investigation of safe drainage of cryogenic fuel vapours;
- for pressure checks of sealings;
- for float valve tests;
- for investigation of inter-tank heat exchanging processes;
- for tests of drainage devices;
- for operation of the whole cryogenic fuel system including the engine.
- technology for large-size cryogenic fuel tank tests at high pressure levels;
- methods for fatigue tests with accelerated temperature/mechanical load cycles.

A number of principal problems regarding the fuel system fire/explosion protection were solved, among them:

- cryogenic tank fire/explosion safety;
- fireproofing system of engine nacelle;
- fire/explosion safety of compartments containing cryogenic equipment.

The following particular components were also developed:

- thermosealed cryogenic couplings;
- cryogenic fuel pipes;
- cryogenic pipe interconnections;
- cryogenic in-flight fuel jettisoning device.

8.4 PROBLEMS TO BE SOLVED IN CRYOGENIC FUEL SYSTEM OPERATION, ENGINE SUPPLY AND REFUELLING

The experience in cryogenic fuel use on the TU-155 aircraft showed that layout and operation of the cryogenic fuel system and its components are far more complex compared to a conventional kerosene system. Taking into account the four main elements – fuel system, engine, APU, ground refuelling facility –, there are three major operational conditions to be considered:

- engine supply;
- APU supply;
- ground refuelling.

8.4.1 Engine supply

A conventional kerosene system has to provide the correct fuel flow to the engine and for temperature and pressure control of the fuel. A cryogenic fuel system, additionally, has to consider the following features:

- Recirculation of fuel during precooling:
  During the engine pump and feed line precooling, intensive fuel evaporation occurs with the formation of large quantities of vapour, which has to be recirculated to the fuel tanks.

- Minimization of precooling time:
  To maintain today's turnaround times of aircraft, it is necessary to minimize the precooling period before engine start-up. This period depends on the tank pump power and the aircraft and engine fuel pipes heat capacities.

- Fuel vapour utilization:
  Due to the fact that the cryogenic fuel temperature is always lower than the ambient temperature, heat flow into tanks, pipes, pumps etc. is always evident. A gas utilization device is to be developed, which feeds vaporized fuel to the engine combustion chamber. The optimization of the total system has to account for this utilization device as well as for insulation efficiency, tank pressure and pump heat flow evolution.

- Pump operation considering cavitation:
  The cryogenic pumps, due to the low fuel density, are rotating at high numbers of revolutions in order to reduce the number of stages. Such pumps require a large cavitation margin, which is related to a power and heat-evolution increase. Tank and engine pump parameters have to be considered within the optimization of the whole system.

8.4.2 APU supply

It has to be decided, whether the APU should be supplied by liquid or vaporized fuel, which, of course, has substantial effect on the layout of the whole fuel system. Due to the low fuel consumption and the heat flow into the pipes, it is not feasible to supply fuel in the liquid condition from the tanks to the APU. The only way would be to feed excessive amounts of liquid fuel to the APU, the major share of which would have to be recirculated to the tanks, resulting in additional heat flow. It is therefore suggested, to supply the APU with vaporized fuel, which has to be considered in the design of the gas utilization system.

8.4.3 Aircraft refuelling

- Vapour release at precooling and tank refuelling:
  Before refuelling the aircraft, the refuelling system has to be precooled, resulting in considerable amounts of evaporated fuel. Moreover, during the refuelling process gaseous fuel has to be succed from the tanks. By joint optimization of the aircraft fuel system and the ground refuelling system it seems to be possible, to reduce considerably, and perhaps to eliminate, the release of vapour into the ground refuelling facility. This will considerably reduce the ground drainage system dimensions and increase the measurement accuracy of refilled fuel.

- Tank heating to eliminate impurities:
  The quantity of dissolved, unwanted impurities in the cryogenic fuel (e.g. oxygen and nitrogen within the LH_2) depends on the temperature. The higher the fuel temperature the more impurities may be dissolved in it. At fuel temperature decrease the dissolved impurities fall out from the solution and form solidified particles. The accumulation of solidified oxygen may lead to explosion and accumulation of nitrogen or mechanical failures of system components. Therefore, regular tank heatings to the temperature of impurities evaporation take place after 5–6 refuellings. Such operational conditions are not acceptable for commercial aircraft. The tank heating during operation must be eliminated by appropriate means to avoid impurities in aircraft tanks as well as in the refuelling system.

- Fuel flow control during refuelling:
  To achieve the same time of pre-flight servicing as for kerosene aircraft, high refuelling rates will be required for the cryo-aircraft. To avoid structural damage by temperature shocks, to increase the fuel metering accuracy and to prevent tank damage in case of fuel level indicator failure, one way may be to reduce the fuel flow at the beginning and towards the end of the refuelling process. The method of fuel flow control during refuelling must be determined for both aircraft and ground facilities.

8.5 CONCLUSIONS AND PROPOSALS FROM THE TU-155 EXPERIENCES

Experience gained during the TU-155 aircraft development and operation draws the following main conclusions:

1. Helium system:
The TU-155 tests showed the impracticability of the installed helium system, used for fuel pipes draining, which required a large quantity of helium on board and a rather complex and heavy control system. All airports would have to be equipped with a high pressure helium refuelling system. For a future series cryo-aircraft, a helium system should be avoided.

2. Development and test of materials and structures:
Aircraft cryogenic components must be light, small-sized and have a long operational life. Due to the necessity of carrying out long-term fatigue tests, it is worthwhile, finding the means for:

- correct choice of light and strong materials, able to operate in a cryogenic medium for long operating times;
- using the available test installations for aircraft and component development;
- advancing development and test of structural elements, especially fatigue tests.

3. Regulations:
There is a great need to start a process to develop certification requirements and regulations and handling rules to ensure safety. This process has to be conducted at least on a European scale and should aim for a worldwide consensus.

4. Servicing, refuelling:
Two major overall requirements for cryo-aircraft pre-flight servicing have to be met:

- simplicity and safety of maintainance;
- analogue procedures and turnaround times to conventional aircraft.

These requirements seem evident but they demand the development of the following:

- refuelling devices and techniques to ensure removal of all impurities before refuelling, in order to avoid periodical tank heating;
- appropriate, possibly automatically operating coupling devices;
- aircraft systems and ground facilities need to have large fuel flows during refuelling.

8.6 OUTLOOK FOR THE FUTURE OF CRYOGENIC AVIATION IN RUSSIA

In the CIS countries, the exhaustion of crude oil reserves, expected by the beginning of the next century, will require transition to alternative fuels earlier than LH_2 can be introduced into aviation on a large scale. The vast resources of natural gas in Russia, estimated at 200–300 trillions of m^3, suggest the use of LNG as an aviation fuel.

The necessity to fulfil the social programme for gasification of many regions of the CIS countries, particularly behind the Urals including Kazakhstan and Middle Asia, will require a solution for the gas transportation/distribution problem, especially for settlements situated far away from large administrative centres. It may be that the CIS gas infrastructure will progress rapidly, allowing the CIS aviation to switch to LNG some 10–15 years earlier than hydrogen operated aircraft will appear in the Western European airlines.

Bearing in mind the ecological situation in Western Europe, it seems futile to widely introduce LNG aircraft, because in a comparatively short time the infrastructure would have to be changed twice, from kerosene to LNG and from LNG to LH_2, which is economically impracticable.

Figure 8.2 shows graphically the forecasted aviation infrastructure change depending on cryogenic fuel introduction. This includes the direct change to LH_2 in Western Europe and the change to LNG in CIS countries with a later switch to hydrogen. The combination of both infrastructures may be highly harmonious. If at Western European airports the hydrogen production could be based upon natural gas (direct decomposition into carbon and hydrogen), the refuelling of LNG aircraft at these airports will not pose major technical problems.

In the CIS countries, at the first stage of the hydrogen technology introduction, a small number of international airports will appear (Moscow, St. Petersburg, Kiev, Minsk, Alma-Ata, Tashkent) where both LNG and LH_2 infrastructures will be available.

As a first generation LNG aircraft, the Tupolev Design Bureau is currently developing the TU-156 M2. The TU-154 well-proven passenger

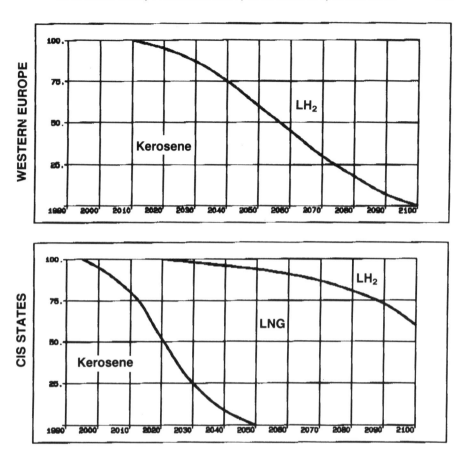

Fig. 8.2 Introduction of cryogenic aviation fuels in Western Europe and in the CIS.

transport was used as the baseline. To convert the selected arrangement with cryogenic fuel tanks on top of the fuselage, the three-engined concept was changed into a two-engined one (Fig. 8.3). All systems not related to the powerplants were retained. This aircraft, could enter service as a

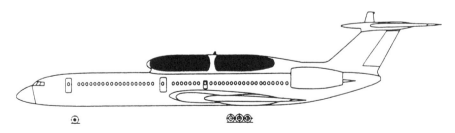

Fig. 8.3 TU-156 M2 passenger aircraft operating with liquid natural gas.

passenger airline at the end of the current decade. It will give extensive practical experience that will be very useful for the further research and development work on the German-Russian Cryoplane project.

9
Aeroengines for Alternative Fuels

V.N. Orlov, V.V. Kharlamov; R. Walther; J. Ziemann
SSSPE 'TRUD', Samara, Russian Federation;
DASA-MTU, Munich, Germany;
Deutsche Aerospace Airbus, Bremen, Germany

ABSTRACT

This paper underpins the need for a change from present day hydrocarbon aviation fuels to alternative cryofuels (liquid hydrogen and liquefied natural gas) and discusses the technical problems arising with the use of those fuels. Engine modifications are necessary to the combustor as well as to fuel system components such as fuel pumps, fuel pipes and control valves. In addition, a heat exchanger will be required for vaporizing and heating the cryogenic-stored fuel. To reduce pollutant emissions from both conventional kerosene fuel or alternative fuels, advanced low-pollution combustor concepts have to be realized. A shift to cryofuels would have immediate benefits since the pollutant emissions into the atmosphere could be reduced drastically.

9.1 INTRODUCTION

Between 1950 and 1970, NASA – its predecessor was NACA – conducted numerous tests on engines for hydrogen combustion [1, 2]. The basic practice was to store the hydrogen cryogenically in tanks, vaporize it with the aid of a heat-exchanger and inject it into the combustor in its gaseous state. It was discovered that burning hydrogen in a gas turbine combustor entailed no major problems. Modifications were only necessary to the fuel injection system, so that it could handle the hydrogen. Hydrogen has a very wide flammability range, which is beneficial to combustor control. Because the fuel is already in its gaseous state prior to injection into the combustor, and because hydrogen burns very rapidly, the required residence time of the fuel/air mixture in the combustor is significantly reduced. This in turn allows a shorter combustion chamber design compared to the conventional kerosene technology. Conditions will be much the same for natural gas combustion, with flammability range and combustion rates in between the corresponding hydrogen and kerosene values.

The Samara State Scientific and Production Enterprise (SSSPE) 'TRUD' began to design engines burning cryogenic fuel at the end of the 1950s. The first experience was gained with the development of bi-propellant (kerosene and liquid hydrogen) liquid rocket engines. A number of liquid rocket engines have been developed in the thrust range from 45 to 150 tonnes featuring unique parameters. The next step was the development of cryogenic aircraft gas turbine engines. Based on this experience, extensive work on cryo-technology at SSSPE 'TRUD' started in the 1970s.

Compared to liquid fuel rocket engines the extended operating life and safety requirements mandatory for aircraft engines created a new challenge. SSSPE 'TRUD' recognized the importance of this work early on, as a contribution to the solution of the global problem of diminishing fossil fuel resources. The urgency of this problem becomes more and more obvious and perceptible. The scientific and technical progress is accompanied inevitably with a steep rise in energy consumption (despite the comprehensive development of energy saving technologies).

Hydrocarbon fuel – the product of oil refining – remains one of the main energy resources. Known world oil resources will be sufficient for the next 40 years based on the predicted consumption rates. For the different regions this period varies and lasts as listed below:

- Former USSR: 13 years
- West European countries: 12 years
- North America: 11 years
- Australia and Asia: 18 years
- Middle East: 108 years
- Latin America: 50 years.

Maximum oil production is predicted for the years 2000–2010 and will then smoothly decrease.

It is expedient to begin prospecting alternative fuels as a substitute for hydrocarbons from decreasing oil resources especially in Europe, North

America, and the former USSR. Moreover, it must be noted that the energy consumption growth could lead to a dramatic ecological change due to the steep increase in the atmospheric emissions provoked by harmful constituents of the combustion products.

We are facing two acute global problems:

- quest of possible use of new alternative fuels;
- development and approval of national energy and transport programmes aimed at significant reduction of pollutant emissions into the atmosphere.

Solving these problems can lead to significant improvements especially in view of the predicted sharp rise in transportation and fuel consumption. It is important to note that the production of the most calorific aeroengine fuel cannot exceed 15% from each ton of crude oil, and even this fraction is optimistic considering the need for other kinds of motor fuel. Cryogenic fuels such as liquid hydrogen (LH_2) and liquefied natural gas (LNG) are considered as excellent alternatives for kerosene. Table 9.1 shows some of the physical-technical characteristics in comparison with kerosene.

Liquefied natural gas consists of:

- methane: 92% ± 6%
- nitrogen: 1.5% ± 1.5%
- ethane: 4% ± 3%
- propane and heavier hydrocarbons: 2.5% ± 2%
- hydrogen sulphide and mercaptan sulphur: < 0.005%

The thermophysical and thermodynamic properties of LNG depend to a great extent on its fractional composition.

LH_2 is the most promising fuel type. It has the highest heating value, reducing the on-board fuel mass requirement to one third. For LNG the difference is less pronounced. LNG has about 15% more calorific energy

Table 9.1 Characteristics of alternative fuels

Fuel	Density kg/m^3	Boiling temperature °C	Lower heating value kcal/kg
Kerosene	780	+220	10,250
Liquid hydrogen	70.8	−253	28,660
Liquid natural gas	422	−161	11,500

than kerosene, which brings fuel mass down to approximately 85% versus kerosene. With regard to the tank volumes needed to carry the reduced fuel mass, however, conditions are clearly less attractive. The specific volume of LH_2 is 11 times that of kerosene and it takes roughly four times the tank volume to store the same amount of energy. For LNG, the tank volume requirement increases by only a factor of 1.6, owing to the smaller specific volume involved.

Alternative fuels afford an advantage when comparing cooling capacities, i.e. the amount of heat that can theoretically be absorbed by the fuels. As it will become apparent, the cooling capacity of LH_2 is 21 times and that of LNG about four times that of kerosene. Present aeroengines operating in subsonic or moderately supersonic regimes have little, if any, use for such enormous cooling capacities. But the situation will be clearly different for advanced airbreathing orbital transports, where the cooling capacity of cryogenic fuels, especially LH_2, is an important factor. Owing to the flight Mach numbers of orbital transports in the hypersonic range, thermal loads on engine components are so tremendous that they would not survive without help from the cooling capacity of cryogenic fuel [3].

9.2 EFFECT ON AIRFRAME, ENGINE AND INFRASTRUCTURE

The physical-technical characteristics of liquid hydrogen and liquefied natural gas inevitably necessitates some structural modifications to the aircraft. The larger fuel tanks lead to increased aircraft dimensions and to some deterioration of the flight characteristics. Also, the technique of fuel storage on board becomes more complicated. It requires heat insulation of the tanks and special on-board systems for provision of the required storage conditions ('cushion' pressurization, pressure monitoring).

However, the analysis made by SSSPE 'TRUD', aircraft manufacturers and research institutes revealed considerable advantages with the use of LH_2 and LNG in aviation. Despite a decrease in the lift-to-drag ratio of 10–18%, the hydrogen powered aircraft shows an advantage in fuel weight (reduced by 64–75% compared to conventional kerosene aircraft), aircraft weight (reduced by 25–51%) and engine thrust (reduced by 12–49%), as shown in Fig. 9.1. The lower values apply to subsonic, the higher values to supersonic aircraft. The lower aircraft gross weight leads to reduced wing loads and dimensions. For supersonic aircraft it reduces the intensity of sonic boom effects on the ground. In addition, less stringent requirements to runway length and surfaces result in lower capital costs for airport construction. The lighter aircraft also leads to lower initial costs. Liquid hydrogen with its

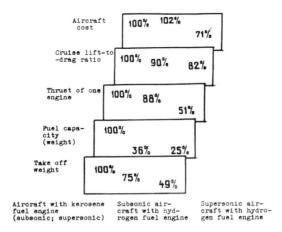

Fig. 9.1 Data comparison for kerosene and hydrogen aircraft.

extremely high cooling capacity and high heating value can even provide a considerable thermodynamic engine cycle improvement.

Figure 9.2 shows a parameter comparison of gas turbine bypass engines fuelled with kerosene and liquid hydrogen at cruise condition. It can be seen that with an increase of bypass ratio and compressor pressure ratio (at constant turbine inlet gas temperature) the LH_2 engine features a reduced specific fuel consumption. For the kerosene engine the relative specific fuel consumption doesn't change considerably at high pressure and bypass ratios. The reasons are now discussed.

For kerosene engines the fuel efficiency can be improved by an increase of both the thermodynamic cycle efficiency and the propulsive efficiency. The specific fuel consumption can be reduced by 1.5–2.0% at a compressor pressure ratio of 30 even at a turbine inlet temperature of 1500 K. However, this may be lost again since it is difficult to obtain a high turbo compressor efficiency and because of the high air offtakes required for turbine cooling. On the other hand, the cooling capacity of LH_2 fuel can be used for turbine cooling via an intermediate heat carrier without the need for air offtakes. It provides either a 10–13% decrease in specific fuel consumption or a 5–6% reduction of the engine dimensions. Thus, the use of LH_2 allows an engine operation at higher compressor pressure ratios and thereby improves the fuel efficiency. It should be mentioned that especially hypersonic aircraft cannot come true without the exploitation of liquid hydrogen. Their severe thermal problems necessitate a fuel with an extreme cooling capacity.

Hydrogen is a renewable energy source. Problems related to hydrogen production in large quantities as well as storage seem solvable. Investigations made in Russia and other countries showed that LH_2 fuel features less

142 AEROENGINES FOR ALTERNATIVE FUELS

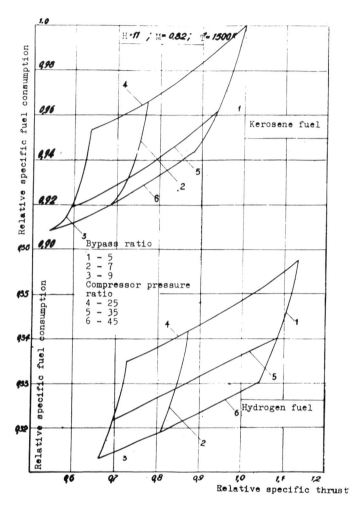

Fig. 9.2 Engine operation with kerosene and hydrogen – characteristic parameters.

of a fire hazard risk than kerosene (when equally spilled) because of its rapid evaporation. Service requirements for liquid hydrogen are comparable to those for kerosene fuel.

It can be concluded that liquid hydrogen deserves more attention as an alternative for hydrocarbon aviation fuel. Its reduction potential for harmful emissions into the atmosphere will be discussed later.

Liquefied natural gas as an alternative fuel has been studied as well. In addition, the possibility of using oil gas has been analysed. Liquefied oil gas consisting of propane and butane has a boiling temperature in the range of −8 to −22 °C and a density of 2 kg/m³ (gaseous). It can be obtained from oil refineries or gas and oil mining fields. It is seldom used in aviation, and there mainly in helicopter engines operated in oil mining/refinery fields. Liquefied natural gas might be a more attractive aeroengine fuel than kerosene considering its world wide resources, its heating value, cost, and reduction potential of pollutant emissions.

The known world resources of natural gas are sufficient for about 58 years based on predicted consumption rates. For the different regions this period may vary:

- Former USSR: 55 years
- West European countries: 34 years
- North America: 14 years
- Australia and Asia: 57 years
- Latin America: 70 years.

The forecast predicts an increased gas production up to the year 2030. It can be seen that a number of world regions can rely on natural gas as a major energy source. The techniques necessary for the production of liquefied natural gas are already developed, and in case of a wide LNG exploitation its costs may fall below those of kerosene.

Compared to kerosene fuelled aircraft, LNG derivations have a substantial advantage despite some deficiencies in flight/technical characteristics, e.g. the IL-86 aircraft powered by two NK-62 M engines burning LNG may have:

- a reduction of aircraft take-off weight by 25.4 tonnes (from 206 to 180.4 tonnes) at constant range,
- increase of range by 600 km at unchanged take-off weight,
- a kerosene fuel saving of 38.1 tonnes per flight (with 19.5 tonnes of LNG usage) without change of range.

Natural gas is distributed via a widely spread net of state and county pipelines and is readily available in the vicinity of many airports.

9.3 CRYOGENIC ENGINE DESIGN AND OPERATION

The operation of cryofuel powered aircraft is feasible only with an adapted infrastructure to provide fuelling at the airports. This includes facilities for production and storage of liquid cryofuel (LH_2 or LNG) and means of transport as well as fuelling stations. Special training is compulsory for both ground personnel and flight crews.

Coming to cryoengine design, we considered the fact of undeveloped infrastructure during the first stage of introduction into service in most regions. Therefore, it is advisable to develop 'two fuel' engines using both cryofuel and kerosene.

Originally, the SSSPE 'TRUD' design team began with the development of an LH_2 engine as a derivative of a conventional one. The series NK-8-2 U bypass engine with 10,500 kg thrust, installed on TU-154 aircraft, was selected as the basic engine. While converting this engine for the new fuel it was decided not to change the gas flow path, the load carrying support structure, or the shaft arrangement. This simplified the design task and allowed us to focus on activities like the selection of an engine pneumatic system, engine control and governor systems, cryofuel pump and its drive, as well as the study of various liquid hydrogen vaporization and combustion processes. The kerosene fuel feed system has been maintained while the cryosystem has been designed around it. The combustion chamber had to be modified to work on both fuels.

The thermal-physical properties of LH_2 as well as established safety standards (fire and explosion proof) determine the following engine configuration requirements:

- all hydrogen units must be attached to the upper external engine casing and especially instrumented for fire/explosion warning, and equipped with air and inert gas purging systems in case of formation of dangerous hydrogen concentrations;
- the main hydrogen units including pipes should be insulated;
- the number of detachable connectors between pipes should be minimized.

The most critical problems identified for the cryosystem are:

- the development of the fuel pump which must operate in a wide range of flow rates;
- the provision of pump operation at low inlet pressure in the required flowrate range (anti cavitation feature);
- the provision of dynamic stability of the fuel system at all engine operation modes;
- the development of a cryofuel combustion chamber providing stable operation.

The engine pneumatic system consists of a combination of units providing fuel supply functions and engine control. This comprises functions like refuelling or purging to remove air from pipes before engine start up.

The various fuel feed system arrangements can be classified in either open or closed loop systems. In the first case (open loop) the fuel pump is driven by a turbine supplied with compressor bled air. The air is then vented. In the second case (closed loop) this turbine is supplied with hydrogen. The closed loop system is more energy efficient, however, it requires the development of pneumatic system components as well as a direct LH_2 flowrate control which is quite complicated.

At the first stage of work the open loop fuel feed system has been selected because it allows the development of separate pneumatic system (PS) components and a comparatively simple pump speed control logic. Furthermore, it was decided to keep the basic kerosene fuel feed system and to design a new combustion chamber for both kerosene and LH_2 fuel operation. The engine control system and PS study was performed considering the combustion chamber operating on both fuels. A single control logic for both operational modes has been selected. The control program has been established using a single setting unit for both fuel feed systems. The kerosene pump has been chosen as the setting unit. In addition a selector valve has been implemented into the engine control system to provide the following functions:

- kerosene operation: fuel supply to the combustion chamber (open valve)
- LH_2 operation: kerosene fuel to the combustor shut off (closed valve) and reverse flow provision into the tank.

For both fuels, engine operation is controlled by the throttle lever position which determines the control pump setting for the chosen high pressure compressor (HPC) rotor speed. In case of operation on LH_2 the kerosene reverse flowrate into the tank corresponds to the throttle lever position. The LH_2 is fed to the combustor at a flowrate required for the selected HPC rotor speed. The hydrogen pump performance (speed, flowrate, pressure) is thereby determined by the pressure signal proportional to the kerosene flowrate. The hydrogen pump speed is varied by compressor air offtake changes. The engine control system and PS are shown in Fig. 9.3.

For the hydrogen engine design, suitable cryogenic components and configurations were thoroughly analysed. The hydrogen pump must be capable of providing a wide range of flowrates during engine operation from start-up to take-off rating. For the type selection, both displacement pumps (gear, plunger, etc.) and dynamic pumps (centrifugal, axial, etc.) were

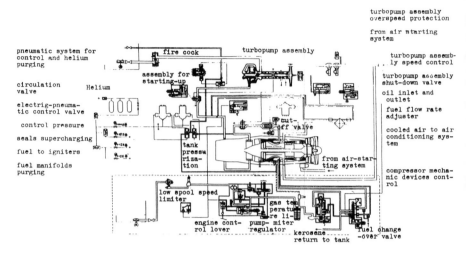

Fig. 9.3 Fuel supply and control system.

analysed. A screw-centrifugal pump has been chosen since sufficient experience on rocket engines with this type has been gained. Moreover, those pumps feature high reliability and efficiency. They feature no friction pairs, and require low torque allowing the employment of a turbine drive type what is considered as the optimum. The pump unit layout is shown in Fig. 9.4.

A number of experiments with the demonstrator engine revealed a hydrogen temperature effect on the circumferential and radial temperature pattern aft of the combustion chamber and on the rotor speed stability. Those effects were monitored at hydrogen temperatures below 60 K, i.e. in the range of a considerable change of its density and viscosity with the temperature. None of these effects have been noticed at higher temperatures. Therefore, a heat-exchanger-vaporizer has been implemented in

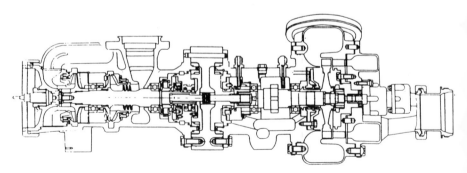

Fig. 9.4 Turbopump assembly.

the fuel supply system for heating the hydrogen before its injection into the combustion chamber.

The selection of the heat-exchanger location has been a major problem. The exhaust gas heat-exchanger, however, has been identified as the most favourable arrangement. It features the required overall dimensions to provide the necessary fuel heating at all engine operating conditions. For the selection of the heat-exchanger type various designs have been investigated experimentally (tubular, plate heat-exchangers and those with ribbed tubes). The multi-tube ring heat-exchanger has been identified as the best configuration considering its weight benefit and simplicity.

One of the main reasons for the introduction of cryogenic fuels in aviation is the environmental protection issue. Various studies indicated a significant reduction potential for harmful emissions when kerosene is substituted by LH_2 or LNG.

At conventional kerosene combustion, the combustor is fed with a low sulphur (containing less than 0.5% by weight) hydrocarbon fuel. The pollutants generated are CO, NO_x, SO_x and unburned hydrocarbons (UHC). The mass of pollutants emitted varies widely with the engine operating condition. In off-design operation, an aeroengine predominantly emits CO and UHC, while at full load and cruise condition mainly NO_x is formed.

The use of hydrogen simplifies the emission problem as products of combustion do not include hydrocarbons, carbon monoxide, or impurities such as sulfur or carbon. It also eliminates benzpyrene which is the most toxic substance formed at the combustion process of kerosene. Concerning the emissions of nitric oxides, hydrogen offers a high potential for a reduction due to its favourable combustion characteristics. This, however, necessitates a redesign of the conventional gas turbine combuster. The unique physical and chemical properties of hydrogen differ considerably from those of kerosene. At stoichiometric fuel/air mixture, the hydrogen flame is 2505 K compared to a kerosene temperature of 2335 K. The hydrogen flame propagation speed is an order higher than that of kerosene, and its range of stable burning conditions is much wider.

To take full advantage of these properties, the combustion process must be tailored to accommodate extremely lean fuel/air mixtures within the primary combustion zone. A lean mixture leads to low combustion temperature which results in a significant reduction of the NO_x production rate, since NO_x formation depends very strongly on the temperature. Lean mixtures can be provided by an increase of the air mass flow entering the bulk head of the combustion chamber from typically 20-30% to 60-70% of the total air flow.

Liquefied natural gas leads to an emission reduction as well. It features a lower flame temperature (2287 K) than kerosene and thus lower NO_x

concentration in the exhaust gas emission. Studies at TRUD indicated a 30% reduction compared to kerosene. The exhaust gas contents also have a lower CO concentration and does not include carbon particles and unburnt fuel in form of aerosols, aldehydes or odorants. However, a redesign of the conventional combustion chamber is necessary for LNG operation. Like hydrogen, LNG is injected into the combustion chamber in a gaseous state. It thereby provides the potential of a high degree of mixture homogeneity which – in a lean mixture – prevents local stoichiometric pockets with peak temperatures and increased NO_x formation (the fuel atomization process necessary for kerosene combustion inevitably includes the probability of fuel droplet deposition on flame tube walls which limits the completeness of chemical reactions within the combustion zone and adversely effects the exhaust gas emission).

During the first development stage it was decided to use a combustor featuring a large number of fuel nozzles, traditional for our engines. This arrangement guarantees favourable burning conditions for both liquid and gaseous hydrogen, as shown in Fig. 9.5.

Some of the nozzle modules were left unchanged for kerosene injection, others (70 out of a total of 139) have been modified for hydrogen. The tests revealed no changes in the main combustion characteristics (combustion efficiency, circumferential and radial temperature pattern, burning stability), see Fig. 9.6. Thus, the large number of nozzles easily permitted a two-fuel combustor operation. A single-fuel combustion chamber for hydrogen is considered as just a simple derivation of this combustor.

During the research and development phase various test rigs have been employed. A full-scale demonstrator engine developed the NK-88 experimental engine operating on LH_2 and kerosene fuel, Fig. 9.7. The bulk of bench and flight tests was performed at operation on LH_2 fuel.

The NK-88 engine has been installed on the TU-155 experimental aircraft designed by Tupolev ANTK, and the engine flight test programme began in

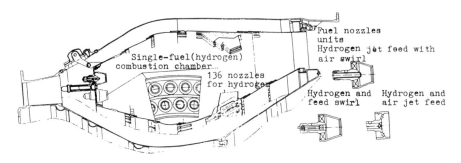

Fig. 9.5 NK-88 engine combustion chamber.

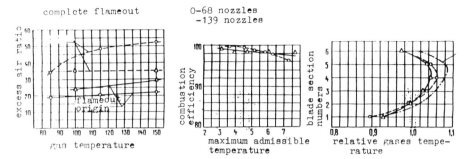

Fig. 9.6 Single fuel and dual fuel combustion chamber characteristics.

April 1988. These tests have proven the viability and rightness of the chosen design. However, all work on the hydrogen engine had been interrupted for the following reasons:

- the present high costs of LH_2 (a tonne of LH_2 cost several times more than a tonne of kerosene);
- the current LH_2 production methods require complex and energy consuming processes;
- LH_2 aircraft fuelling requires the simultaneous development of suitable airport infrastructure.

Considering the exhaustion of oil reserves, and expecting the development of low cost LH_2 production methods in the near future, we predict a widespread use of hydrogen as a universal fuel in the 21st century. Keeping this in mind, SSSPE TRUD and ANTK Tupolev began to think about the intermediate use of liquefied natural gas as an alternative fuel for aircraft in the early 1980s.

The experience gained this far with the LH_2 work has been beneficial for this task. Experiments with the NK-88 engine modified for LNG operation started in 1986. All hydrogen fuel feed system elements basically have been kept while switching to LNG. Some changes have been made at the engine fuel feed turbopump and control system. Furthermore, the surface area of the heat-exchanger has been increased. Bench tests have proven the technical solutions and the feasibility of LNG operation.

The TU-155 flying test bed modified for LNG operation made its first flight in January 1989. A series of test flights has been performed thereafter. A demonstration flight from Moscow to Nice (France) via Bratislava has been conducted in October 1989 on occasion of the 9th International Natural Gas Conference. The TU-155 made another international flight

AEROENGINES FOR ALTERNATIVE FUELS

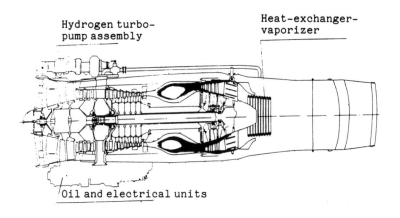

Parameter	Regime Take-off			Cruise	Fuel – liquid hydrogen
Flight altitude, km	0	0	0	11	
Mach number	0	0	0.26	0.8	
Ambient temperature, °C	+15	+30	+30	−56.5	
Ambient air pressure, kgf/sm^2	1.0332	1.0332	1.0332	0.2914	
Thrust, kgf	10500	10165	8500	2200	
Specific fuel consumption, kg/kgf·h	0.22	–	–	0.296	
Reduced air flowrate kg/s	236	–	–	231	Fan diameter 1355 mm.
Bypass ratio	1.06			1.09	
Pressure ratio	11			10.5	Engine weight 2300 Kg
Turbine inlet gas temperature, K	1140	1178	1176	956	

Fig. 9.7 NK-88 bypass engine.

from Moscow to Hannover (Germany) via Minsk during the International Air Show in May 1990. A third flight has been from Moscow to Berlin for demonstration purpose during the International Natural Gas Congress in July 1991.

The performed work including bench tests and flight tests enabled us to identify all major problems concerning the cryogenic technology. All of these problems are considered to be solvable.

9.4 PRESENT ACTIVITIES AND OUTLOOK

The next step is to design an airline transport. ANTK Tupolev (Aviation Scientific and Technical Complex) began to develop the TU-156 passenger/cargo airliner for Aeroflot. This aircraft will have three NK-89 engines operating on LNG and kerosene and will enter into service at the end of this decade. The NK-89 engine has been designed by the SSSPE TRUD design bureau headed by General Designer N.D. Kuznetsov. Assembly of the first engine has already been completed. Experimental testing is ongoing.

While cryogenic aeroengines so far were derived from existing kerosene engines, the large-scale switch to alternative cryogenic fuels will require the development of new engines starting from the scratch.

An international working group has been established, formed by aviation companies in Russia, the USA and Germany. The purpose is to study the feasibility of cryofuel aircraft and engines from a technical and operational point of view, with an emphasis on LH_2 as the most prospective long-term aviation fuel. The three principal domains – aircraft, engine and infrastructure – have been considered.

Three aeroengine companies – SSSPE TRUD (Russia), Pratt & Whitney (Canada) and MTU (Germany) – are engaged in a joint engine study group. The goal is to evaluate the required work as well as the development risk. The study includes a number of work packages for the most significant modifications such as the combustion chamber, engine fuel pump and evaporator.

Presenting a low-emission combustor concept, Fig. 9.8 examplarily depicts a radially staged lean combustor operating on hydrogen. Local temperature

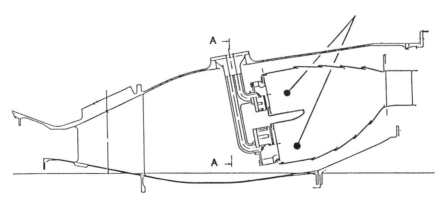

Fig. 9.8 Radially staged lean combustor for low-pollution H_2 combustion.

peaks, which would lead to an increased NO_x formation, are largely avoided by a large number (64) of fuel injection nozzles providing a comparatively homogeneous lean mixture. Furthermore, the radial fuel staging allows low-emission operation also at off-design conditions at high combustion stability and efficiency.

A major problem, however, arises at the development of suitable cryogenic fuel pumps. Fuel pumps for aeroengine gas turbines must operate at a wide range of mass flows. Moreover, present safety and lifetime requirements are challenging. Times between overhaul are required to be long. The existing database from rocket propulsion engineering is barely applicable, since rocket flight missions are typically short, and the need for controlling fuel flow does not generally exist.

A suitable pump to deliver cryogenic fuels would be the centrifugal type [4, 5]. This type is characterized by relatively moderate mechanical friction and accompanying wear on the impellers. Its only big problem is the life of the shaft bearings and seals operating in a cryogenic temperature environment. There is also the need to avoid cavitation at the impeller.

Figure 9.9 illustrates a two-stage air turbine-driven centrifugal pump to deliver cryogenic hydrogen. High pump efficiency requires high rotational speed and a large number of pump stages. However, high speeds severely limit bearing life, and a large number of stages adds to the weight of the pump. The two-stage centrifugal pump for speeds around 45,000 rpm for LH_2 represents a trade-off between efficiency on the one hand and life plus weight on the other. The fuel pump can be driven mechanically, electrically or pneumatically. Figure 9.9 shows a two-stage axial-flow turbine as a drive

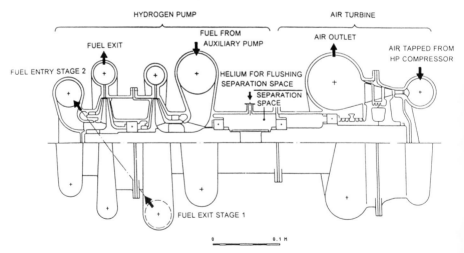

Fig. 9.9 Two-stage centrifugal pump with air turbine drive for H_2 supply.

option with infinitely variable speed. The turbine in turn can be driven with bleed air tapped from the engine compressor.

A further important fuel system component is a cryogenic heat-exchanger to vaporize and sufficiently heat cryogenic fuels before injection into the engine combustor. Figure 9.10 shows an exhaust gas heat-exchanger installed between the LP turbine and the exhaust nozzle.

Another option for the heat-exchanger installation is in between the HP compressor and the combustor (Fig. 9.11). This configuration seems attractive especially for retrofitting existing kerosene engines. With the overall length of the H_2 (or natural gas) combustor being shorter than that of a kerosene type, the resultant gap between the HP compressor and the combustor can conveniently be utilized for this purpose.

Prime requirements for the heat exchanger, apart from the ability to sufficiently heat the cryogenic fuels over the entire range of engine loads,

Fig. 9.10 Exhaust gas heat-exchanger.

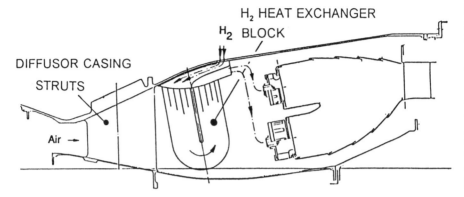

Fig. 9.11 Profile tube heat-exchanger integrated between HP compressor and combustion chamber.

are safety and adequate component life. This is of special significance in view of the high static and dynamic structural loads and of the prevailing high temperature gradients. Icing is another problem, but can be avoided by judiciously selecting the design parameters.

As mentioned earlier, the tremendous cooling capacity of cryogenic fuels can hardly be utilized in present turbojet engines operating in subsonic or moderately supersonic regimes. One potential use for that capacity may be in the cooling down of air for the turbine blades cooling. A further potential use may be offered by the cooling of engine oil needed to lubricate and cool bearings. In conventional kerosene engines the fuel is already being used for engine oil cooling.

9.5 SUMMARY

1. The worldwide increase in fuel consumption immediately necessitates – at least partly – the substitution of hydrocarbon fuel to save non-renewable oil reserves.
2. The increasing atmospheric emissions burden is becoming a more and more important environmental issue. The demand for principle decisions concerning environmental protection is evident.
3. Liquid hydrogen and liquefied natural gas are considered as excellent alternatives for fossil hydrocarbon fuels.
4. The physical and chemical properties of these fuels guarantee reduced pollutant emission rates. ANTK Tupolev demonstrated the feasibility to introduce cryofuels in aviation. Engine bench test results and flight tests

with the TU-155 aircraft proved the adopted technical solutions for liquid hydrogen and LNG operation.
5. The large-scale introduction of clean alternative cryofuels in aviation must be the next stage in this work.

9.6 REFERENCES

1 Esgar, J.B. State of Technology on Hydrogen Fueled Gas Turbine Engines, NASA TM X-71561, 1974.

2 Kaufmann, H.R. High-Altitude Performance Investigation of J65-B-3 Turbojet Engine with both JP-4 and Gaseous Hydrogen Fuels, NACA RM E57A11, 1957.

3 Krammer P., Schwabb R.R. Engine Technologies for Future Spaceplanes, MTU Focus, 1/1992.

4 Baerst, C.F., Riple, J.C. Preliminary Studies of a Turbofan Engine and Fuel System for Use with Liquid Hydrogen, *DGLR/DFVLR International Symposium on Hydrogen in Air Transportation*, Stuttgart, 1979.

5 Sosounov, V., Orlov, V. Experimental Turbofan Using Liquid Hydrogen and Liquid Natural Gas as a Fuel, *AIAA/SAE/ASME/ASEE 26th Joint Propulsion Conference*, Orlando, Fl. July 1990.

10
Alternative Fuels in Aviation

H.W. Pohl
Deutsche Aerospace Airbus, Hamburg, Germany

10.1 SUMMARY OF PREVIOUS STUDIES AND EXPERIMENTAL WORK ON ALTERNATIVE FUELS IN AVIATION

Apart from early applications of hydrogen in balloons and airships, projects to use hydrogen as an aviation fuel began in the United States some 40 years ago.

In 1955, a report of the NACA-Lewis Flight Propulsion Laboratory was published in which the potential of liquid hydrogen for use in both subsonic and supersonic aircraft was explored.

As a result of this study, an experimental programme was initiated to demonstrate the feasibility of burning hydrogen in a turbojet engine at high altitude. A US Air Force B57 twin-engine medium bomber was modified as shown in Fig. 10.1 and first flown in 1956. Liquid hydrogen (LH_2) was carried in a tank located under the left wing tip. Gaseous helium was carried in a tank of similar size and shape under the right wing tip for use as a pressurant.

Throughout the flight test programme, performance of the engine at altitude was found to be exceptionally smooth and reliable, and in conformity with expectations; no operational safety problems with the hydrogen fuel system were encountered.

At the same time, Lockheed Aircraft Corporation initiated the design of a hydrogen-operated Mach 2.5 reconnaissance airplane. This project, however, was dropped due to the logistics problem posed by the fuel.

In 1973, a series of studies sponsored by the National Aeronautics and Space Administration (NASA) were begun to explore the potential of hydrogen in aircraft at greater depth.

The early studies performed for NASA started with analysis of supersonic

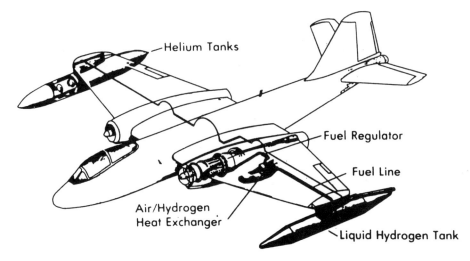

Fig. 10.1 B-57 aircraft used in NACA hydrogen tests.

aircraft. Conceptual design studies of subsonic aircraft started the following year.

Essentially, the objectives of the studies in both speed regimes were identical. They were to assess the feasibility of using hydrogen fuel in commercial transport aircraft, to determine its advantages and/or disadvantages relative to conventional Jet A fuel, to identify problems and technological requirements associated with use of LH_2, and to outline a plan for development of required technology.

In order to focus the Lockheed effort on work related to the aircraft side of the problem, it was mandated that the LH_2 was to be available in storage at the airport. While the aircraft studies were underway at Lockheed, NASA funded other studies to investigate processes, energy requirements, and costs involved in manufacture, transmission, and liquefaction of hydrogen. This latter work was done by the Institute of Gas Technology, Chicago, and the Linde Division of Union Carbide Corporation, Tonawanda, NY.

A general arrangement of the LH_2 fuelled, supersonic transport aircraft design which resulted from the preliminary studies performed by Lockheed California Co. for NASA-Ames Research Center is shown in Fig. 10.2. The aircraft was designed to carry 234 passengers 4200 nm, at a cruise speed of Mach 2.7.

The passengers are located in a double-deck arrangement in the central section of the fuselage. LH_2 fuel is contained in tanks forward and aft of the passenger compartment.

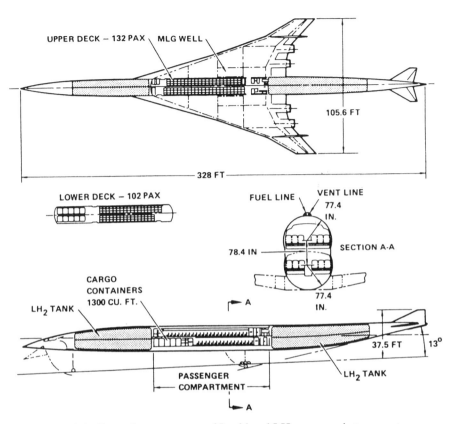

Fig. 10.2 General arrangement of Lockheed LH$_2$ supersonic transport.

Figure 10.3 is a general arrangement of a preferred configuration of LH$_2$ fuelled, 400-passenger subsonic aircraft designed to fly 5500 nm at a cruise speed of Mach 0.85. The general layout of the passenger compartment relative to the fuel tanks is similar to that described previously for the supersonic transport aircraft, i.e. the passengers are located in the central portion of the fuselage in a double-deck arrangement with the fuel tanks located fore and aft.

Similar design studies were carried out for subsonic transports operating on liquid methane.

For both supersonic and subsonic aircraft the fuel tanks were designed as integral tanks with a foam insulation of 5–6 in.

In Europe, VFW, MBB and Handley-Page were involved in alternative fuel projects (Fig. 10.4). These early conceptual studies were important in that they served to establish some of the basic tenets of LH$_2$ fuelled aircraft.

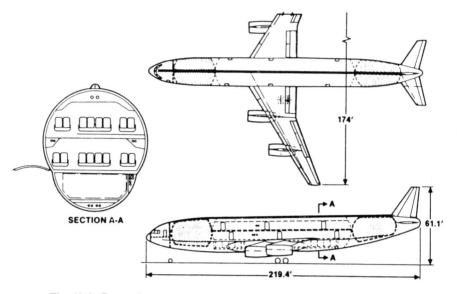

Fig. 10.3 General arrangement of Lockheed LH$_2$ subsonic transport.

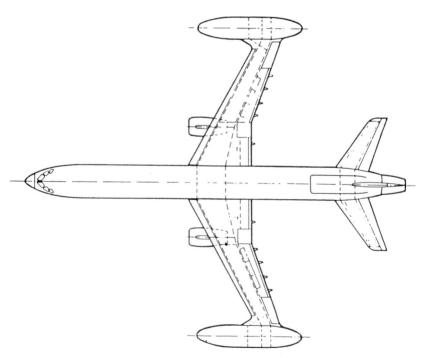

Fig. 10.4 VFW design for LH$_2$ subsonic transport.

Also, they helped to bring the subject of hydrogen to the attention of the aircraft and air transport industries, showing that there is a viable alternative to conventional jet fuel and that it offers significant advantages in many areas.

However, the main driver to initiate the activities on alternative fuels was the oil price crises in the 1970s. When oil prices declined again during the 1980s, the studies were largely discontinued.

10.2 CURRENT STUDIES AT DEUTSCHE AEROSPACE AIRBUS

Research work on alternative fuels at DA began in 1988 and is supported by the German Ministry of Economics and the Euro-Québec Hydro-Hydrogen Pilot Project.

Coordinated by DASA-Airbus, 15 partner companies are involved in the studies (Fig. 10.5), including the Russian companies Tupolev and TRUD, which are operating the TU 155 experimental aircraft.

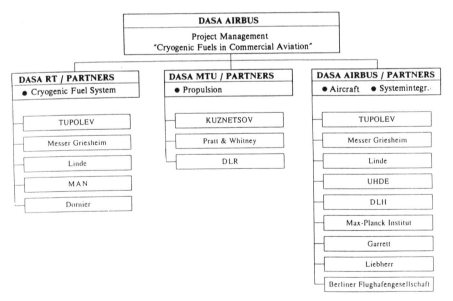

Fig. 10.5 Partner companies involved in the Cryoplane project.

10.2.1 Basic assumptions and aircraft requirements

One of the major topics considered in the studies is to develop scenarios for the introduction of an alternative fuel into aviation.

Obviously, the transition to a new fuel cannot be accomplished on a world-wide scale simultaneously, because substantial financial efforts are required for aircraft and engine development as well as infrastructure build-up. It is likely that the introduction of a new fuel will start in an economically and technologically highly developed region like USA, Japan or Europe.

Looking at Europe as the starting point, some major requirements for a first generation alternative fuel aircraft can be identified. To cover all intra-European routes, a design range of 2700 nm is required. Due to the relatively long development periods of transport aircraft (some ten years for a conventional aircraft plus another five to ten years technology phase in case of an alternative fuel aircraft), the first production aircraft could enter into service between 2005 and 2010. By this time, the intra-European traffic will mostly be accomplished by a fleet of 400–500 wide-body aircraft. Hence, the category of medium-range wide-body aircraft was selected for the configuration studies.

To limit the development costs, the first generation alternative fuel aircraft should be a modification of an existing conventional type of transport airplanes rather than an all-new design. Hence, the maximum take-off weight is predetermined by the baseline aircraft.

The infrastructure for the alternative fuel has to be built up according to the development and entry into service of the aircraft. After substitution of the total fleet of 400–500 aircraft, which will require some 20 years, infrastructure has to be available at about 70 European airports. In the case of hydrogen, a yearly production of 2 Mio t of LH_2 will be required, which compares to some 7300 t of today's yearly production.

10.2.2 Aircraft configuration studies

An Airbus A310 was selected as baseline aircraft for the configuration studies (Fig. 10.6). This wide-body aircraft has a maximum take-off weight of 150 t seating 243 passengers in an all-tourist layout. It is powered by 2 PW-4152 engines of 52,000 lb static thrust each. Configuration studies were carried out with liquid hydrogen (LH_2) and liquid natural gas (LNG) as alternative fuels. The weights of both cryogenic fuels are lower than the weight of kerosene assuming the same energy content:

- LH_2:   0.36 × weight of kerosene;
- LNG: 0.85 × weight of kerosene.

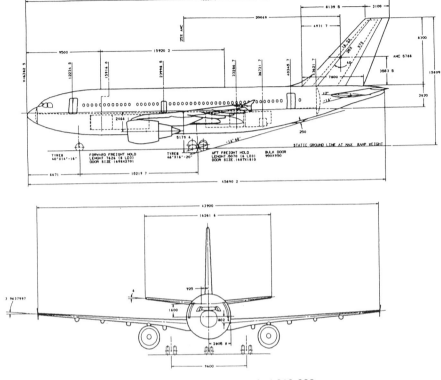

Fig. 10.6 Baseline aircraft A310-300.

On the other hand, fuel volumes are larger for same energy contents:

- LH_2:   4 × volume of kerosene;
- LNG: 1.6 × volume of kerosene.

Furthermore, fuel storage at cryogenic temperatures requires insulated pressurized tanks, which are to be designed most favourable in a spherical or cylindrical shape, i.e. conventional wing tanks are not applicable.

The aircraft/tank configurations selected for the design studies comprised tanks under the wings, on top of the fuselage and in the fuselage.

The weight benefit of the cryogenic fuels is transferred into additional passenger capacity by stretching the fuselage, thus maintaining the baseline take-off weight of 150 t.

Although the configuration studies considered both LH_2 and LNG, all partners concluded – as a result of a feasibility study – that LH_2 is to be preferred as future aviation fuel, because:

- NG, like crude oil, is a fossil resource. The remaining reserves in terms of years are comparable to those of crude oil.
- Compared to kerosene, the CO_2 emission of NG is only 25% lower.
- The main component of NG – methane – is a greenhouse gas itself, some 20 times more effective than CO_2.
- Due to the higher fuel weight, the stretch potential of LNG aircraft is relatively low.

However, LNG may be a useful interim solution for CIS-countries, where natural gas resources fairly exceed the crude oil reserves and kerosene is scarcely available in remote areas.

LH_2 tanks below the wing (Fig. 10.7):

Similar to the engine installation, the tanks in this configuration are mounted to the wing by pylons. A fuel weight of 16,400 kg including reserves is required for the design range, corresponding to a volume of 230 m^3 (115 m^3 per tank). The fuselage of the baseline A310 is stretched by 22 frames, increasing the passenger capacity by 87 to 330.

The dynamic characteristics of the wing (torsion, flutter) are severely affected by the LH_2 tanks. It seems unlikely, that the wing structure can be modified to withstand the dynamic loads. Another penalty of this configuration is a substantial deterioration of low-speed performances, as parts of the high-lift systems are covered by the tanks.

LH_2 tanks in the fuselage (Fig. 10.8):

The LH_2 is carried by two tanks, installed in the front and rear section of the fuselage outside the pressurized cabin area. Due to the required tank volume, the fuselage length is significantly increased, which leaves no stretch potential for the passenger cabin. The fuselage length in this case is limited by the rotation angle required for take-off and landing.

The seating capacity is the same as on the baseline aircraft, resulting in a significant operating cost penalty of some 20% compared to the other designs. However, this configuration might be a favourable solution for an all-new design where fuselage cross-section is a variable parameter.

LH_2 tanks on top of the fuselage (Fig. 10.9):

The fuel is carried by four tanks installed along the fuselage upper side, only omitting the area affected by a potential engine burst. The vertical tail is

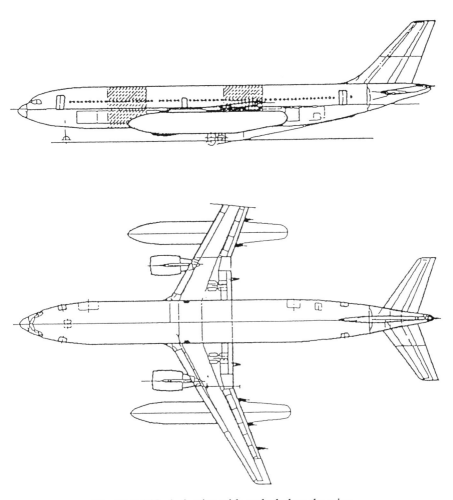

Fig. 10.7 LH$_2$ derivative with tanks below the wing.

shifted upwards to avoid covering by the rear tank fairing. The baseline fuselage is stretched by 18 frames, corresponding to a passenger capacity of 310. The fuel weight including reserves amounts to 15,500 kg.

To select the most favourable solution, performances, operating costs, handling and safety aspects were analysed. Considering the basic assumptions of these studies (modification of an existing conventional aircraft, using the fuel weight benefit for passenger capacity increase) the tanks-on-fuselage configuration turns out to be the best solution.

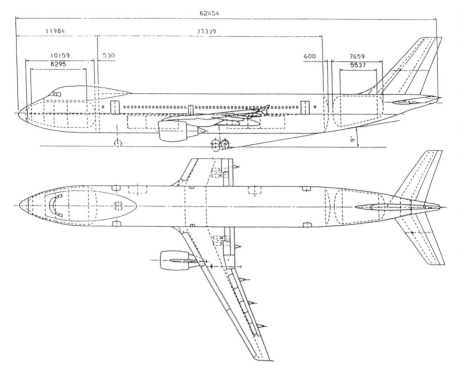

Fig. 10.8 LH$_2$ derivative with tanks in the fuselage.

10.2.3 Cryogenic fuel system

Major modifications are necessary for the fuel system of the LH$_2$ aircraft which has to fulfil the following main requirements:

- A fuel quantity of 15,500 kg including reserves is required for the design range, corresponding to a volume of 218 m^3. Considering additional volume due to APU operation, system components installed in the tanks, expansion of fuel etc. the gross volume amounts to 240 m^3.

- The insulation of the tanks has to be designed for an overnight stop (12 h) without evaporation of hydrogen.

- The turnaround time between flights shall not exceed the usual turnaround times of kerosene aircraft. Thus, the refuelling of the LH$_2$ aircraft must be possible at the gate within some 30 min.

- Due to safety aspects, no fuel system component must be installed

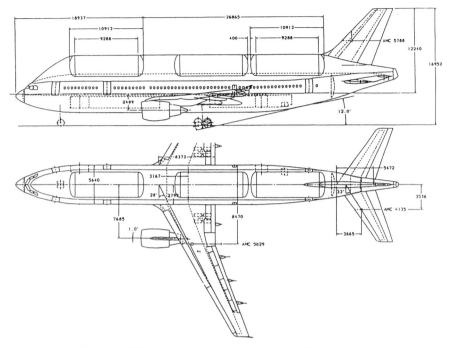

Fig. 10.9 LH$_2$ derivative with tanks on top of the fuselage.

within the pressurized cabin area (avoidance of explosive hydrogen/air mixtures).

- The operational life of system components has to correspond to the aircraft's operational life (e.g. 40,000 flight-h for a short/medium range airplane).

Although LH$_2$ has been used as a propellant in space systems for a long time, fuel system components are not applicable due to the short operational life. Substantial technological development is necessary to meet the extreme requirements regarding reliability and lifetime in commercial aviation, e.g.

- Material characteristics at cryogenic temperatures have to be verified for tank structures and pipes. Carbon-fibre reinforced plastics and a newly developed Al-Li alloy are considered as candidates.
- Insulation materials are to be analysed to meet the 12 h requirement (overnight stop without evaporation). Super-vacuum insulation and a new mineral fibre recently developed in Russia are favourite candidates.

- Bearings and sealings with operational life of several 10,000 h at cryogenic temperatures have to be developed for the fuel pumps and valves.
- Control systems and sensors to detect hydrogen leakages are to be developed.

10.2.4 Economical aspects of LH_2 aircraft

Direct operating costs (DOC) comprising capital costs, insurance, crew and maintenance costs, fees and fuel costs were analysed for both the LH_2 design and a reference kerosene aircraft. To obtain a preliminary estimation of the price of an LH_2 aircraft, non-recurring and recurring costs for the necessary modifications of the baseline aircraft were evaluated, based on a total production of 450 aircraft. Adding these modification costs to the actual purchase price of the A310 results in a minimum price for an LH_2 production aircraft. Furthermore, the DOC calculations were based on the following data:

- average stage length of 1500 nm;
- kerosene price: 0.47 DM/kg (0.70 \$/gal);
- LH_2 price: 6.00 DM/kg (e.g. electrolysis by hydropower).

A comparison of operating costs for the LH_2 and the reference kerosene aircraft is presented in Fig. 10.10, showing the DOC of the LH_2-design at nearly 160% of the reference value. The share of fuel costs in case of the LH_2 aircraft is in the order of 50%.

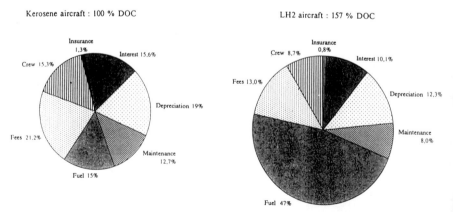

Fig. 10.10 Comparison of operating costs, kerosene and LH_2 aircraft.

Looking at future price developments for both kerosene and LH$_2$, opposite trends can be expected (Fig. 10.11):

- increase of kerosene price by environmental taxes or abolishing of tax exemption, long-term increase by crude oil shortage;
- reduction of LH$_2$ price by improved production methods, use of excess power of electrical power plants, advances in solar power generation and mass production (significant effect of production quantity on price).

As a rule of thumb, the LH$_2$ aircraft may be considered competitive to the kerosene aircraft, if the energy related fuel prices are equivalent.

10.2.5 Ecological aspects

A comparison of the various emissions of both kerosene and hydrogen is given by Fig. 10.12. The primary products of kerosene combustion are CO$_2$ and H$_2$O (at 3.1 kg/kg fuel and 1.2 kg/kg fuel resp.). Beside these, carbon monoxide, unburned hydrocarbons, nitric oxides, sulfur oxides and soot are emitted to a smaller amount.

In case of hydrogen, the greenhouse gas CO$_2$ is completely avoided, leaving only water as the primary combustion product. At the same energy

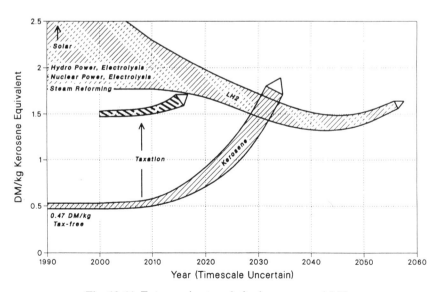

Fig. 10.11 Future price trends for kerosene and LH$_2$.

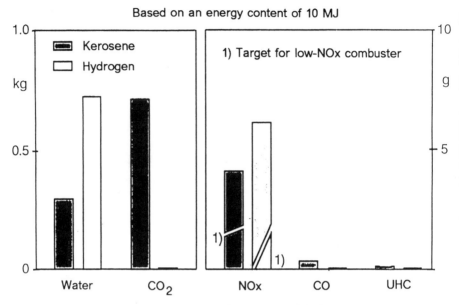

Fig. 10.12 Emissions of kerosene and hydrogen.

contents, the mass of water is about 2.6 times as high as the water emission of kerosene. The only by-products of hydrogen combustion are nitric oxides.

Nitric oxides:

Beside CO_2, nitric oxides are the most crucial emissions of aeroengines due to their effect on ozone concentration (depletion of stratospheric ozone, formation of tropospheric ozone). Since the introduction of jet engines into commercial aviation, substantial effort has been applied to reduce fuel consumption, mainly by increasing pressure ratio and turbine inlet temperature. As the formation of NO_x is directly depending on these parameters, the relative amount of NO_x (per unit of fuel burned) has been increased.

The ICAO in December 1991 agreed to aim for 20% lower emission levels and the EC is suggesting to reduce NO_x from jet engines by even 40%.

New concepts for combustion chambers aiming for NO_x reduction are currently developed, mostly based on lean or rich combustion, e.g.

- lean premixed prevaporized combustion;
- lean, quick-quench, rich combustion;
- radially staged lean combustion.

Fundamentally, these concepts are applicable to both kerosene and hydrogen. However, in the case of kerosene lean or rich combustion increases the formation of other pollutants (CO, UHC, soot). Furthermore, the lean mixture of the kerosene/air mixture is very restricted due to the narrow flammability range. The flammability range of hydrogen is by a factor of 14 greater than that of kerosene, i.e. the hydrogen/air mixture can be made leaner to a much greater extent and without penalties due to other pollutants. Hence, a greater NO_x reduction potential is to be expected for hydrogen.

Water:

Water emissions from jet engines may occur as water vapour or contrails (ice clouds).

Figure 10.13 represents the greenhouse effect of water vapour, which is

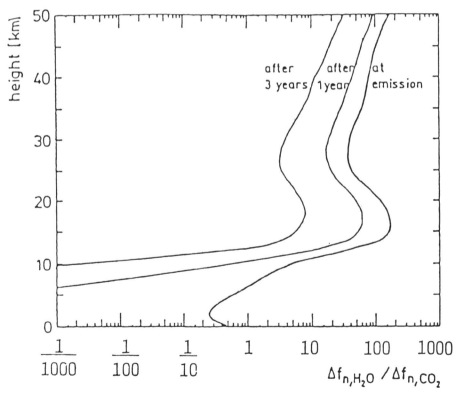

Fig. 10.13 Greenhouse effect of H_2O related to CO_2 for equivalent number of molecules.

dependent on altitude, related to CO_2 for an equivalent number of molecules. At altitudes above 6 km the effect of water vapour exceeds the CO_2 effect ('at emission'). However, the residence times of H_2O and CO_2 in the atmosphere are different by several orders of magnitude:

- CO_2: more than 100 years (independent of altitude);
- H_2O: 3–4 days at ground level, increasing to 0.5–1 year in stratosphere.

If aircraft operations are limited to the troposphere (8–9 km in arctic, 13–15 km in tropic regions) no significant greenhouse effect of water vapour is to be expected. Due to the short residence time, the amount of water vapour emitted by human activities is negligible compared to the atmosphere's natural water content.

CO_2 emissions, however, are accumulating because of the extremely long residence time. If emission is continued at today's rate, the CO_2 from human activities will reach the same amount as the atmosphere's natural CO_2 content within 100 years.

Figure 10.14 compares the greenhouse effects of alternative fuels at varying flight levels for mid-latitude atmospheric conditions, accounting for

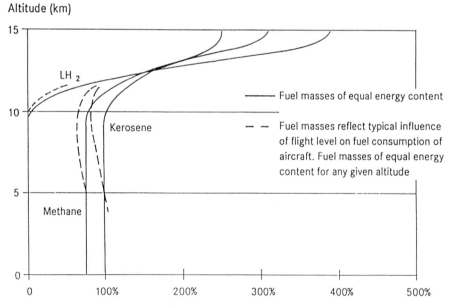

Fig. 10.14 Greenhouse effect of various fuels.

the residence times mentioned above. Comparing masses of equal energy content (full lines), it is found that:

- Below 10 km the contribution of water vapour is negligible. As the effect of CO_2 is independent of emission height, the effect of burning kerosene is constant (= 100%) up to about 10 km. Methane produces somewhat less CO_2, hence causes less greenhouse effect. LH_2 produces only water vapour, hence the greenhouse effect is very close to zero.

- Above 10 km, water vapour becomes more and more important, and dominates beyond 12 km. Hence, the effect of kerosene increases (now CO_2 *and* H_2O being effective), methane and LH_2 cause greater greenhouse effects than kerosene.

In general, aircraft use less fuel the higher they fly, due to reduced drag at lower air densities. If this trend is allowed for, the dashed lines of Fig. 10.14 result. It can be seen, that the current flight levels 31,000/33,000/35,000 ft (i.e. around 11 km) cause the minimum greenhouse effect (by gaseous combustion products).

Water emissions may contribute to the greenhouse effect when forming contrails. As shown in Fig. 10.15, contrail formation is dependent on ambient temperature and altitude. At a typical cruising level of 10 km, no contrails are to be expected at temperatures above −42 °C. But, contrails

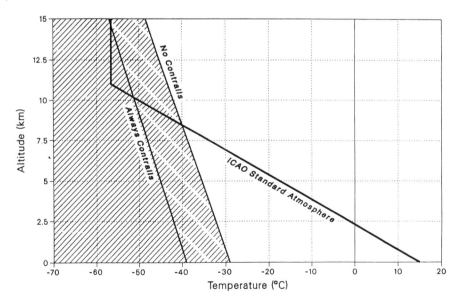

Fig. 10.15 Formation of contrails.

174 ALTERNATIVE FUELS IN AVIATION

will occur at temperatures below $-50\,°C$. If the temperature is between these limits, contrail formation depends on the actual condition of the atmosphere. For example, contrails frequently occur within the warm sector of low pressure areas.

Depending on the actual meteorological conditions, the flight altitude may have to be reduced to avoid contrails, resulting in higher fuel consumption and thus higher operating costs. The effect of altitude variations on the operating costs of the hydrogen aircraft has been analysed and resulted in only moderate cost increases (1 to 3%).

It has to be reminded that the atmospheric impacts of water emissions are fundamentally the same for both kerosene and hydrogen aircraft, the difference being only the amount of water per unit of fuel burned.

Summarizing the environmental considerations, the hydrogen aircraft has a significant advantage versus the kerosene aircraft, due to:

- absence of greenhouse gas CO_2;
- absence of all by-products except NO_x;
- higher NO_x reduction potential.

10.3 CRITICAL COMPONENTS AND R&TD REQUIREMENTS

R&D activities are required in the field of aircraft configuration. Future cryo-aircraft designs, instead of being derived from a kerosene aircraft, are to be optimized for the use of cryogenic fuels from the scratch. Problems due to large fuel volumes will particularly concern long-range LH_2 aircraft. This may lead to unconventional solutions requiring in-depth studies of aerodynamics, loads and structures, flight mechanics.

Synergy effects with other aircraft systems, eg. using the cooling potential of cryofuels, are to be investigated.

Further research and development mainly concerns the components of the cryogenic fuel system, which is to be completely redesigned.

Structural materials for tanks and pipes:

- candidate materials: new Al-Li-alloy and CFRP (carbon-fibre reinforced plastics);
- analysis of embrittlement and permeability in LH_2;
- fatigue strength at cryogenic temperatures;

- possibility of chemical milling and weldability of Al-Li-alloy;
- production techniques, especially for CFRP tanks.

Insulation:
- candidates: supervacuum and mineral fibre insulation;
- analysis of heat flow vs. thickness;
- load carrying/transferring characteristics of fibres;
- research on inner insulations (significant weight benefit expected).

Pumps and valves:
- analysis and tests of bearings for LH_2 pumps with $3 \times 40,000$ h of operational life;
- sealings for pumps and valves.

Control system:
- development of H_2 sensors to detect fuel leakages.

Operational aspects:
- use of evaporated fuel, e.g. if ground times exceed 12 h;
- development of a fuelling/defuelling valve with reliable protection against pollution of cryofuel during fuelling processes.

Safety analysis:
- large-scale tests on spilling high quantities of cryofuel;
- investigation of deflagration/detonation transition.

11
Alternative Fuels in Ground Transportation

R. Buchner

Daimler-Benz Central Research, Stuttgart, Germany

11.1 INTRODUCTION

A drastic reduction of emission limits for large cities/conurbations is to be expected. This applies particularly to the emissions of motor vehicles. Concepts envisaging the use of low or zero emission vehicles in inner cities are being discussed (e.g. exhaust gas legislation in California and other US states).

The considerations are concentrating on vehicles with:

- electric propulsion system (battery/electric motor);
- hydrogen propulsion (H_2 storage tank/internal combustion engine);
- hydrogen/electric propulsion (H_2 storage tank/fuel cell/electric motor).

Numerous research and development programs have clearly demonstrated that hydrogen can be burned in an internal combustion engine resulting in nearly zero emissions of atmospheric pollutants. Futhermore, unlike battery-powered total electric vehicles, automobile performance with hydrogen fuel is similar to that with gasoline.

11.2 DESCRIPTION OF THE HYDROGEN PROPULSION CONCEPT WITH INTERNAL COMBUSTION ENGINES

11.2.1 Vehicle concept

To drive a vehicle with hydrogen, modified spark ignition engines can be used and operated with hydrogen (external and internal mixture formation). The volume-related output and the consumption of hydrogen engines are theoretically between the relevant values for diesel and gasoline engines.

When burning hydrogen in the engine, no CO, CO_2, CH or lead emissions can be traced in the exhaust gas. The power of a hydrogen-fuelled engine with external carburetion is approximately 20% less than that of a gasoline engine with the same volume. H_2 engines can be operated with extremely lean mixtures and therefore with extremely low NO_x emissions. The reduced output which results from the lean operation can be compensated by supercharging (Fig. 11.1).

Since hydrogen-fuelled internal combustion engines emit practically no pollutants in lean operation (only lubricating oil emissions) they are close to electric motors in terms of environmental compatibility (Fig. 11.2).

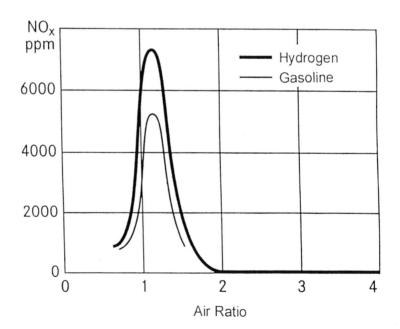

Fig. 11.1 NO_x formation depending on fuel/air ratio.

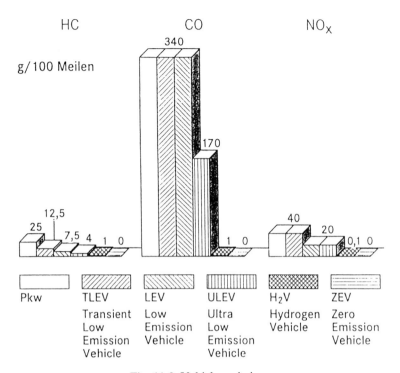

Fig. 11.2 Vehicle emissions.

11.2.2 On-board hydrogen storage

Cars with electric or hydrogen propulsion will not be able to simply substitute today's cars, the reason being that – with regard to the volume and the weight – batteries and hydrogen storage tanks store much less energy than a gasoline tank. The higher efficiency of electric propulsion as compared to internal combustion engines can only partly make up for the unfavourable battery storage densities (Fig. 11.3).

The concern for safe hydrogen storage has been solved by utilizing metal hydride storage tanks. In this approach, hydrogen is bonded to metals under moderate pressure and released with the application of heat. Heat energy is readily available from the engine by circulating coolant (water) or the hydrogen fuel exhaust product (steam) through the metal hydride tank. Since heat has to be put into the storage tank when driving to release the hydrogen, e.g. from the air inside the passenger compartment, the heat medium is cooled down at the same time. Thus, it is possible to have a CFC-free air conditioner on H_2 vehicles. Figure 11.4 shows the specific data and the configuration of an on-board hydride storage tank.

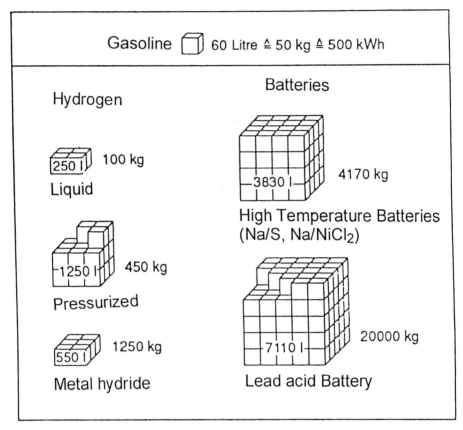

Fig. 11.3 Energy densities.

Hydrogen storage on passenger cars/vans is already proven technology in the form of metal hydrides (the equivalent of five litres gasoline is stored in a 130 kg hydride tank). Advantages of hydride storage are:

- higher storage density than batteries;
- short charging times (ten minutes);
- high degree of safety.

High-pressure storage tanks (300–400 bar) are currently being investigated at DBAG in a pilot project for use in a bus (HYPASSE project).

A liquid hydrogen (LH_2) tank weighs only about two to four times as much as a gasoline tank for the same range. The cost of the LH_2 tank is

Mobile Hydride Storage Tank ("Ideal Battery")

Spec. data

Energy density: 330 - 400 Wh/kg; 770 Wh/l
Power density: 2000 W/kg
Cost 250 - 300 DM/kWh
Lifetime: > 300.000 km
(without capacity loss)
Recyling: o.k.
FCKW-free air conditioner (3 - 5 kW)

Module data

Weight: 120 kg
Volume: 52 Litres
Energy content: 1.2 kg H_2 = 40 kWh 0 4,5 l gasoline
Range: ~ 60 km CDC, HDC
Refuelling time: 10 min. (100 %)
Cost: ~ 10.000 DM
Cost/ 100 km: < 3 DM

Fig. 11.4 Hydride storage tank.

relatively low, while its service life is high (> ten years). This represents a particularly economic storage solution. The boil-off rates are approximately 2% LH_2/day. The remaining questions regarding the technology and safety of an on-board LH_2 tank are currently being investigated. Medium-term solutions have been envisaged.

11.2.3 Hydrogen vehicles

H_2 vehicles with hydride storage tanks have been successfully tested in a pilot scheme (Berlin), at customers' operation. The individual components are fit for standard production, however, they have generally not been optimized (Fig. 11.5).

Dual fuel operation with gasoline and H_2 is technically feasible and offers the following advantages:

- usual range outside of cities (gasoline operation);
- nearly zero emission vehicles in cities (H_2 operation);

182 ALTERNATIVE FUELS IN GROUND TRANSPORTATION

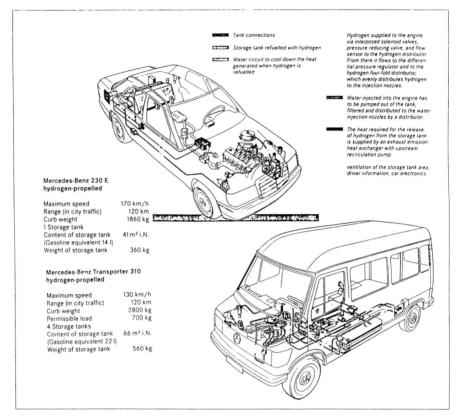

Fig. 11.5 Hydrogen vehicles with hydride storage.

- only one vehicle required, no need to change the vehicle;
- engine waste heat for heating;
- use of hydride storage tank for air conditioning (CFC-free), also during gasoline operation.

Disadvantages:

- The efficiency of the engine can only be optimized for hydrogen. However, the increased gasoline consumption is only a few percent (with respect to optimized gasoline operation).
- Two carburetion systems required.
- Two tanks.

11.2.4 Availability

Nearly zero emission vehicles, which are to be introduced into the market in 1998, must build on technologies which are already widely available today:

- different storage tanks (hydride tank, high pressure tank, liquid hydrogen tank);
- components of the hydrogen regulation system (some still have to be designed to suit automotive requirements).

However, essential components of LH_2 technology are not yet available including:

- injection pumps for internal carburetion;
- feed pumps for liquid hydrogen with sufficiently long service lives.

11.3 STATE OF HYDROGEN VEHICLES

11.3.1 Daimler Benz AG (DBAG)

In 1972, DBAG began working with hydrogen-fuelled internal combustion engines and hydride tanks with the aim of finding a low-emission alternative to electric drives with batteries, and of avoiding the risks associated with pressurized gas containers and liquid hydrogen tanks. This work was concluded in 1988 with successful customer trials of hydrogen-fuelled vehicles with hydride storage tanks. The cars in Berlin had the following features:

- modified serial engines with a volume of 2.8 litres for gasoline and H_2 mixed operation (not dual-fuel operation);
- engine power 120 kW at 5500/min;
- external carburetion;
- hydride tank: 330 Wh/kg.

The pick-up trucks in Berlin had the following features:

- modified serial engines with a volume of 2.3 litres;
- engine power 75 kW at 5600/min;

- external carburetion;
- water injection to avoid backfiring, knocking, or self-ignition and to minimize NO_x (no serial solution, will be replaced by lean operation);
- hydride tank: 330 Wh/kg.

Customer trials with a fleet of five automobiles and five pick-up trucks (Fig. 11.5) were carried out in Berlin (Berlin fleet) in the period from October 1984 to March 1988. The hydride tanks were charged with extremely pure hydrogen obtained from the Berlin city gas system at a conventional gasoline station where hydrogen pumps had been installed.
Results:

- acceptable vehicle operation corresponding to trial status;
- vehicle fuel consumption approximately similar to that of vehicles in gasoline operation (excess weight of hydride tank must be balanced against better consumption of H_2 engine);
- running time of the hydride tanks (up to present):

  —individual tanks > 100,000 km (approximately 1500 filling cycles)

  —total mileage of the fleet vehicles > 700,000 km

  —reduction of the tank capacity of 10% after 1500 cycles (100,000 km).

  —tank can be thermally (approximately 300 °C) regenerated to approximately 95% of the original capacity
- simple filling procedure;
- filling times: ten minutes at 30–50 bar;
- cold start possible.

11.3.2 BMW

Work on hydrogen engines using liquid hydrogen as an alternative to gasoline and diesel commenced in 1980 at BMW, comprising some test vehicles from the series 5 and series 7 with the following features:

- LH_2 tank (developed by Linde Co., Messer Griesheim);
- engine with external and internal carburetion;
- power: 80 kW.

11.3.3 Mazda

Experiments with hydrogen Wankel engines and hydride tanks as an alternative to electric vehicles have been carried out since 1984 on city vehicles with the following features:

- hydride tank (developed by Nippon Steel Co.), approximately 400 Wh/kg;
- Wankel engine: power approximately 30 kW.

11.3.4 Other automobile manufacturers

- H_2 engine development by: Renault, Peugot, Citroen, Fiat, Volvo;
- cooperation MAN/BMW in the manufacture of local busses with hydrogen engines.

11.4 COSTS

11.4.1 Fixed costs

In practice, although the fixed costs of a hydrogen-driven vehicle are higher than those of a gasoline-operated one, the difference only amounts to the costs of the hydrogen tank. Specific tank costs amount to:

- hydride tank: 300 DM/kWh;
- high-pressure tank (not usable in passenger cars; no reliable information has been provided by the manufacturer to date);
- LH_2 tank: 30 DM/kWh.

11.4.2 Variable costs

The variable costs are principally dependent on the hydrogen costs. The hydrogen costs (GH_2), given in litres of gasoline equivalent, range between:

- 0.60 DM (natural gas steam reforming plant) undistributed ex factory;
- 3.00 DM (small electrolysis unit with electricity costs of 0.16 DM/kWh).

The operating costs can be reduced:

- By series production of the hydride tank. This would reduce costs by around 30% (< 200 DM/kWh tank costs).
- By reducing the installed tank capacity. Since a hydride tank can be filled up in ten minutes, it can be matched to meet the requirements of a daily journey – principally in the city – of around 40 km, including a safety margin. This reduces the tank costs by approximately 50% and the operating costs by about 15%.
- By reducing the vehicle weight and the driving power. This leads to a clear reduction of the hydrogen consumption, the tank weight and the investment costs for the storage tank.

11.5 IDENTIFYING REASONS FOR THE INTRODUCTION OF HYDROGEN TECHNOLOGY AND OBSTACLES STILL TO BE OVERCOME

11.5.1 Reasons for hydrogen introduction

- The necessity of reducing pollutant emissions in cities: laws, e.g. in the USA, based on that planned for 1998 in California.

11.5.2 Obstacles

- Extensive regulations dealing with the certification of H_2 vehicles and H_2 gasoline stations (particularly with regard to cities, underground car parks) are not yet in place.
- H_2 supply/infrastructure is not yet comprehensively available in cities. The availability of hydrogen at competitive prices in easily-accessible gasoline stations is the most important prerequisite for the introduction of hydrogen-powered vehicles. It is thus absolutely vital to develop an infrastructure for supplying these vehicles with hydrogen fuel.

11.6 EVALUATION OF THE APPLICATION POTENTIAL

On the one hand, the application potential is related to the advantages offered by a hydrogen-driven vehicle:

- the very low emission levels, which at $\Lambda < 2$ are close to zero emission;
- in the case of hydride tanks:

  —a small hazard potential

  —can be filled up within ten minutes

  —loss-free storable energy

  —the possibility of operating a heating system using heat dissipated by the engine

  —the possibility of obtaining CFC-free cooling from the tank during discharge.

On the other hand, potential demand is also dependent on the extent to which the general public is prepared to accept H_2 as a fuel and to which a comprehensive H_2 supply is guaranteed.

The application potential is dependent on administrative and legal factors. The introduction of hydrogen-powered vehicles will only be realized in comparatively large numbers and not gradually.

The gaseous hydrogen produced locally could be stored safely in metal hydrides during the initial phases of the introduction of hydrogen as a vehicle fuel.

The LH_2 technology and its components have been known for a long time from space technology (NASA) and tried and tested.

Because of remaining technical questions (filling stations, delivery pumps etc.) and safety issues (boil-off losses, oxygen condensation, approval for conurbations etc.) comprehensive application of LH_2 vehicles cannot be expected for the near future (until approximately the year 2010).

LH_2 technology will be used in the medium and long-term in motor vehicles once practically zero emission vehicles are required outside of conurbations, too.

12
Phase II and Phase III.0 of the 100 MW Euro-Québec Hydro-hydrogen Pilot Project (EQHHPP)

J. Gretz
European Commission, Joint Research Centre, Ispra (VA) Italy

12.1 INTRODUCTION

The concept of a hydrogen-based, clean, renewable energy system, conceived by the Joint Research Centre Ispra of the European Commission, is currently investigated by European and Canadian Industries, coordinated by the JRC-Ispra of the European Commission and the Government of Québec.

The 100 MW pilot project is to demonstrate the provision of clean and renewable primary energy in the form of already available hydro-electricity from Québec converted via electrolysis into hydrogen and shipped to Europe, where it is stored and used in different ways: electricity/heat cogeneration, vehicle and aviation propulsion, steel fabrication and hydrogen enrichment of natural gas for use in industry and households.

Phase II, the detailed system definition, investigated costs of the electrolytic hydrogen, produced with hydropower which would be available at 2 cents$_{ECU}$/kWh, shipped to and stored in a European port, of 14.8 cents$_{ECU}$/kWh in the form of liquid hydrogen.

The present Phase III.0 is a hydrogen demonstration programme on the utilization of hydrogen in the fields of vehicle and aviation propulsion, steel fabrication and advanced techniques of liquid hydrogen storage. This phase also involves detailed studies of safety measures and codes, along with socio-economic studies on the comparison of hydrogen with conventional fuels.

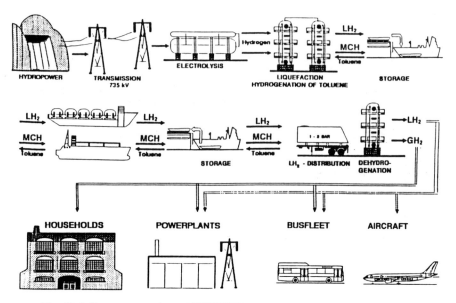

Fig. 12.1 System overview of EQHHPP concept investigated in Phase II.

12.2 THE CONCEPT

The 100 MW (el) pilot project [1] is to demonstrate the provision of clean and renewable primary energy in the form of Canadian hydropower. This is converted via electrolysis into hydrogen and shipped to Europe, where it is stored and used in different ways: electricity/heat cogeneration, vehicle and aviation propulsion and hydrogen enrichment of natural gas for the use in industry/households (Fig. 12.1).

For reasons of thermodynamic properties, availability of technology and end-use, two different modes of vectorization have been investigated namely liquid hydrogen (LH_2) and methylcyclohexane (MCH) in order to have hydrogen in both forms, liquid and gaseous, available for the end-use requirements.

12.3 MILESTONES

The project is to be carried out in the following phases:

- Phase I: assessment, 1986/04/01–1987/03/31

- Phase II: detailed system definition, 1989/01/01–1991/03/31
- Phase II Suppl. Task Programme: additional R&D programmes in Europe, 1990–1992
- Phase III.01: pre-approval activities in Europe, 1991
- Phase III.02: hydrogen demonstration programme, 1992–1994
- Phase III.03: hydrogen demonstration programme (European extension), 1993–1998
- Phase III: detailed engineering and specifications, duration 1–2 years
- Phase IV: construction, duration 4–5 years

12.4 PHASE II OF THE EQHHPP

In Phase II of the project [3], the Phase I results have been updated, cost calculations refined, environmental analyses undertaken and questions of safety and regulations investigated.

In the following the results of these investigations are given, based on the EQHHPP's concept described in the publication of Phase I: 'The 100 MW Euro-Québec Hydro-hydrogen Pilot Project', International Journal of Hydrogen Energy, Vol. 15, No. 6, pp. 419–424, 1990.

12.4.1 The partners

The industrial partners involved in the project are (alphabetically):

Air Liquide Canada (CND), AEG AG (D), Ansaldo Ricerche (I), Autobus MCI (CND), Blohm & Voss AG (D), BMW AG (D), CONOC Continental Contractors (D), Daimler Benz AG (D), DECHEMA (D), Ecole Polytechnique (CND), Electrolyser Inc. (CND), FEDNAV Ltd. (CND), Fenco Lavalin Inc. (CND), Fraunhofer Institut für Systemtechnik und Innovationsforschung (D), Gaz Métropolitain (CND), GERAD (CND), Germanischer Lloyd AG (D), Hamburger Hochbahn AG (D), Hamburgische Elektrizitäts – Werke AG (D), Hamburger Gaswerke GmbH (D), Holinger GmbH (D), Hamburgische Gesellschaft für Wirtschaftsförderung GmbH (D), Hydrogen Industry Council (CND), Hydrogen Systems NV (B), Institut Francais du Pétrole (F), Joint Research Centre Ispra of the Commission of the European Communities (I), L'air Liquide SA (F), Linde AG (D), Messerschmitt – Bölkow – Blohm GmbH (D), Messer Griesheim

GmbH (D), Paul Scherrer Institut (CH), Pratt & Withney (CND), Reederei August Bolten (D), SNC/FW Ltd (CND), Staatliche Materialprüfungsanstalt der Universität Stuttgart (D), STCUM (CND), Technische Hochschule Darmstadt (D), Technische Universität Hamburg – Harburg (D), The LGL Group Ltd (CND), Thyssen – Nordseewerke GmbH (D), Uhde GmbH (D), Union Eéctrica Fenosa SA (E), Universidad da Las Palmas de Gran Canaria (E), VTG – Paktank GmbH (D), Université Concordia (CND), Université Laval (CND).

12.4.2 Project management

Phase II of the project was managed by a Joint Management Group (JMG) consisting of the Ludwig-Bölkow-Stiftung (LBS), Ottobrunn, Germany, and Hydro Québec (HQ), Montreal, Québec. For the subsequent Phases III.0, III and IV, Demonstration Projects and Realization, a new Arbeitsgemeinschaft for the European side has been established, an ARGE consisting of the Ludwig-Bölkow Systemtechnik (LBSt), Ottobrunn, Germany, and CONOC Continental Contractors GmbH (CONOC), Hamburg, Germany.

12.4.3 Reference case (LH₂ vector)

As reference case LH_2 was considered, with a hydroelectric capacity of 100 MW to be converted into gaseous hydrogen via a 74%$_{(uhv)}$ efficient electrolysis technology, operated at 8300 h/yr (95% capacity factor) and feeding a liquefier. Hence, the following data apply:

- hydropower   100   MW
- electrolysis (net)   74   %
- annuity (8% interest, 15 yr pay-back)   11.7   %
- load factor   95   %
- hydrogen delivered in Hamburg   74   MW = 614 GWh/y
- hydrogen transmission efficiency   74   %
- plant investment costs   415   MECU
- specific hydrogen energy costs   14.8   cents$_{ECU}$/kWh

The LH_2 produced is to be transported in one barge container ship carrying five vacuum insulated barge vessels with a content of 3000 m³ LH_2 to

Europe in 17 round trips per year. The overall system efficiency (hydro-electricity to LH_2 delivered in port) is above 50%.

12.4.4 Safety

On 18 December 1990 the BAM (German Federal Institute for Material Research) declared that liquid hydrogen would be no more dangerous than LNG and LPG and that it has no objection to the transport of liquid hydrogen in two G type ships (IMO rules).

12.4.5 Costs

At investment costs of 415 MECU and an annuity of 11.7% (8% interest, 15 year pay back) the feasibility study indicates costs of liquid hydrogen at port in Europe of 0.15 ECU/kWh_{th} or approximately 1.70 ECU per litre of gasoline equivalent at hydro-electricity costs of about 2 $cents_{ECU}$/kWh_e (sensitivity analysis indicates a 19% impact on the product cost by doubling electricity cost).

The specific product costs and their breakdown for the LH_2 vector (reference case) are given in Figs 12.2 and 12.3.

The specific product costs of gaseous hydrogen (GH_2) from the MCH

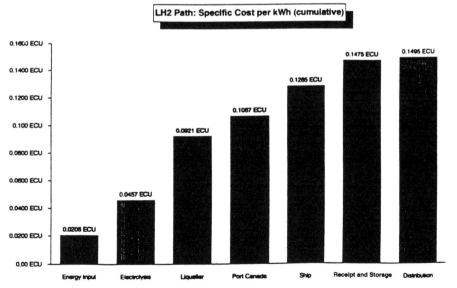

Fig. 12.2 Specific costs in ECU/kWh (LH_2 reference case).

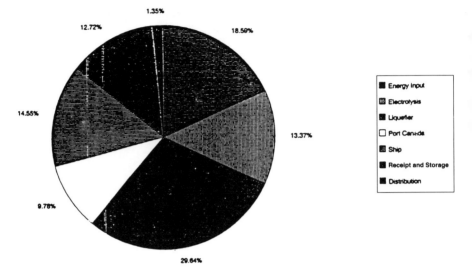

Fig. 12.3 Specific costs distribution (LH_2 reference case).

vector are 12 cents$_{ECU}$/kWh if dehydrogenated with hydrogen, the specific product costs are 15.3 cents$_{ECU}$/kWh. Therewith, the product costs of the MCH vector is higher than those of the LH_2 vector if it is dehydrogenated with clean hydrogen which has to be the case if the product is to be clean in all its steps from production to utilization. Furthermore, the user's profile requires 62 MW LH_2, and 12 MW GH_2. To satisfy this, 62 MW of the GH_2 in case of the MCH vector have to be liquefied by part of the GH_2 arrived at Hamburg, a procedure which brings the product costs of the LH_2 from the MCH vector up to 22 cents$_{ECU}$/kWh.

The investment costs contain the battery limit costs as specified by the industrial partners and the costs for off-sites and auxiliaries. In addition, indirect costs and interest incurred during construction are included. The complete cost figures are therefore higher than the costs of the naked plants at the battery limits.

A standardized calculation for depreciation was used with an interest rate of 8% and a capital payback period of 15 years resulting in a constant annuity of 11.7% (1990 money). The operating costs include a.o. energy consumption and consumables.

12.4.6 Vector selection

Weighting the pros and cons of the LH_2 vector vs. the MCH vector, i.e. considering the advantages of MCH:

- unlimited storage periods,
- transport and storage in existing normal oil product ships and containers,

against its disadvantages:

- the energy intensive dehydrogenization and the liquefaction done at the user's site whereas the energy intensive liquefaction of LH_2 is done with abundant hydropower in Québec,
- MCH as well as toluene are petrochemical products and therewith environmentally less advantageous than LH_2,
- its product (GH_2) is not adapted to the user's profile, about 80% of hydrogen use being in form of LH_2,

the decision was taken to retain LH_2 as vector for the realization of the project.

12.5 ECONOMICS

The hydrogen cost of 14.8–15 cents$_{ECU}$/kWh are pictured in Fig. 12.4 together with the costs of gasoline prices in Europe (average of the 12 EC countries, August 1990) of 8.5 cents$_{ECU}$/kWh which are made up of 3

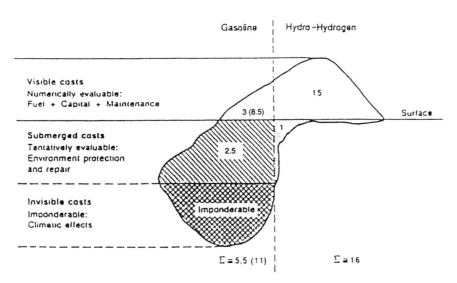

Fig. 12.4 Cost 'Iceberg'.

cents$_{ECU}$/kWh for the crude oil itself, its transportation, refinement, manipulation and distribution, and of 5.5 cents$_{ECU}$/kWh for taxes.

With the above values and including the submerged costs of protection/repair of the damaged environment resulting from the use of fossil fuel (the value of 2.5 cents$_{ECU}$/kWh is the average from various literature sources), the 'fuel cost iceberg' would look as shown in Fig. 12.4, indicating the merits of clean technologies.

With internalization of external costs, hydrogen energy would be about (1.45) 2.9 times higher than hydrocarbon energy costs, not taking into account the imponderable costs of climate effects.

12.6 ENVIRONMENTAL ASPECTS

As it is the case for any energy system, the construction of hydropower plants and the use/management of hydrogen has an impact on the local and the global ecosystem:

- Hydropower: the usually quoted worldwide potential of $15-20 \times 10^3$ TWh/y is only 0.0057% of the hydrological cycle energy and is based on topological, technical and environmental concerns. In any case, careful site selection and adequate measures have to be undertaken in order to minimize the impacts.

- Water vapour: the worldwide water evaporation from the oceans and rivers is $\approx 5 \times 10^{14}$ m^3 per year. If man's total energy consumption of a sustained 11 TW could be brought about by hydrogen, its yearly evaporation would be 2.5×10^{10} m^3 i.e. about 1/20000 of the natural evaporation. Once hydrogen is massively used, local considerations would be obligatory such as in the case for today's wet cooling towers.

- End-use: whereas the combustion of hydrogen does not produce CO_2, CO, SO_2, VOC and particles, it entails emission of water vapour and NO_x.

- Water vapour emissions from aircraft may be harmful since they generate – depending on the cruising altitude and latitude – ice clouds with ensuing greenhouse effects. The problem is of utmost importance and actually under investigation; cruising altitude management is one answer. The formation of NO_x from the atmospheric components N_2 and 0_2 is a function of flame temperature and its duration. Considering the wide flammability range of hydrogen, its combustion can be influenced by the design of the engine so that the NO_x emission can be reduced. Hydrogen is an excellent fuel for fuel cells, its more or less cold combustion will reduce the NO_x emissions even down to zero.

12.7 PHASE III.0

A fund of 8 MECU has been made available by the European Commission in 1991 and 6.7 MECU by European industry as well as 6.3 MECU from the Government of Québec and 2 MECU from Canadian industry to undertake, in parallel to Phases III and IV demonstration projects on the utilization of hydrogen in four fields where the use of hydrogen shows its attractiveness. In Europe, these projects are executed by industry on the basis of contracts following a tender action by the Commission with a cost sharing of at least 50% by the industrial partner:

- Vehicle propulsion/public transport: In Germany, for example, from the overall emissions those resulting from road traffic are: 52% NO_x, 70% CO and 49% VOC.

  A typical cost breakdown for public transport operation in Europe shows that fuel cost contribution to total operation costs is only between 5% and 10%, whereas personnel costs amount to some 67%. Therefore, the introduction of clean, and thus often still more expensive fuels, seems feasible earlier in the public transport sector.

  Three bus projects for public urban transport are presently under way in Europe and shall lead to tested hardware systems. Two buses will be operated with internal combustion engines adapted to gasified liquid hydrogen, one based on a MAN diesel engine (Hydrogen Systems, Belgium) and the other on a MAN gas engine (MAN, Germany). One bus with electric drive and battery storage (Ansaldo, Italy) will be powered by a membrane fuel cell system (from SERE de Nora, Italy). All three projects work on full size city buses carrying liquid hydrogen as fuel in superinsulated tanks onboard (400, 630, 600 litres). Demonstration of the buses on public roads will take place mainly in the years 1995 and 1996. A fourth bus concept (Esamco, Ireland) with a stirling engine, electric drive, battery storage and compressed hydrogen storage for a midi-size bus is not yet active.

  In Québec, three buses to be operated with hythane (a mixture of typically 20% hydrogen and 80% natural gas) will be engineered, fabricated and tested (by Novabus Inc.). The hythane engines will be provided by Europe and from the USA. Hythane is regarded as a very cost effective solution for the introduction of clean fuels in public city buses complying with Californian ultra low emission vehicle requirements.

In Italy, a small public transport boat on the Maggiore Lake will be converted to membrane fuel cell propulsion and LH_2 storage (Ansaldo) and shall be demonstrated by the end of 1995, early in 1996.

- Aviation propulsion: Different combustor nozzle concepts are under investigation and two concepts will be tested in a combustor chamber section of an Airbus jet engine in steady state and transient testing. Goal of this joint activity (DASA Airbus and Pratt & Whitney Canada) is to investigate the potentials for NO_x reduction dependant on flame temperature, residence time and air/fuel ratio and provide design guidelines for later development of a real flight combustor. These activities were started in 1992.

    Hydrogen in aviation would not only reduce carbon and sulphur-related emissions in high altitudes but also on-ground at airports. The emission of higher water vapour amounts (2.5×) compared with kerosene jet engines can be managed by latitude dependant cruising altitude management, presumably at very minor additional costs.

- Steel fabrication: With approximately 2 kg CO_2 emissions per kg steel fabricated with carbon as reductant, the world's steel fabrication contributes over 10% to the world's total anthropogenic CO_2 production. Hydrogen as an excellent and clean reductant has a large potential to reduce CO_2 emissions.

    An iron ore and scrap melter reducer component with two plasma arc injectors of 0.8 MW is built and tested, demonstrating the direct reduction of iron ore fines with hydrogen as reductant yielding crude steel (Kent Steel). The goal is to commence demonstration in a steel plant (preferably at a larger output capacity) as soon as possible and to build a commercially operating plant rapidly. Such an optimized plant of a larger scale shall include additional CO_2 neutral reductants such as forest wastes, for example.

- Cogeneration: Two different cogeneration projects are pursued in this project phase, one on the basis of a 30 kW_e/70 kW_{th} piston engine and the other on the basis of a 200 kW_e phosphoric acid fuel cell. The piston engine cogeneration plant will be operated in conjunction with a stand-alone wind energy converter in Belgium after build-up and testing in 1992/1993 is finished successfully. The fuel cell plant will be tested as basis for later full sized cogeneration plants in the 10 MW range. The fuel cell activity will last from 1992 to 1998.

- Advanced techniques of liquid hydrogen storage: Large-scale model containers will be built and tested, including accident simulation and rupture tests.

In addition the following will be undertaken:

- economic studies related to social costs and the comparison and the use of conventional fuels and of hydrogen and definition of the advantages of hydrogen,
- investigations on the potential of lowering the costs related to the use of hydrogen and to demonstrate that it can be introduced as a clean and safe alternative fuel,
- detailed studies of safety codes and requirements and risk evaluation.

12.8 FUNDING

The accumulated EQHHPP budget for the past and present activities in Phase II Feasibility Study, Phase II Supplementary Task R&D Program Europe, Phase III.0-1 Europe, Phase III.0-2 and Phase III.0-3 Europe amounts to approximately 45 MECU, out of which 16 MECU were covered by participating European industry, 3 MECU were covered by partcipating Canadian industry and 19.7 MECU came from CEC and 6.1 MECU from Québec Government. In Europe, these demonstration projects are executed by industry on the basis of contracts following tender actions by the EC with a cost sharing of at least 50% by industrial partners (Note: 1 MECU = 1 million ECU/1 ECU equals approximately 1.20 US$, March 1993).

12.9 OUTLOOK

An investigation is under way on the management and organization structure for the realization of the EQHHPP project. A Euro-Québec Hydro-hydrogen joint undertaking as a buyer, future owner and operator of the project and a Euro-Québec Hydro-hydrogen Project Management, both being judicial bodies, are foreseen.

Capital costs of the overall system of producing, transforming, storing, transporting and delivering hydrogen would be high and further developments have to be made to render them economically acceptable. Phase III.0

addresses this question of lowering the costs while it also addresses the question of showing the merits of hydrogen as a clean and dependable fuel.

12.10 ACKNOWLEDGEMENT

The present report is largely based on the works undertaken by the representatives of the industrial and scientific partners of Phase II of the EQHHPP. The authors are grateful to all persons involved in this project for the quality of their work and their engagement in the project.

12.11 REFERENCES

1 The 100 MW Euro-Québec Hydro-Hydrogen Pilot Project, J. Gretz, J. Baselt, O. Ullmann, H. Wendt – *International Journal of Hydrogen Energy*, Vol. 15, No. 6, pp 419–424, 1990.

2 *EQHHPP – Phase II Final Report*, Joint Management Group, 1991.

Index

Note: Figures and Tables are indicated by *italic page numbers*

A310 Airbus, 162, *163*
  LH$_2$-fuelled configurations, 164–5, *165*, *166*, *167*
Aeroengines
  alternative fuels used, 127, 140–54
    international study group, 151, *161*
  cryofuelled
    design characteristics, 141, *142*, 143–50
    fuel supply and control systems, 131, 145–6, 152–3, 166–8
    heat-exchanger arrangements, 153–4
    low-emission design, 151–2
Aircraft
  alternative fuels for, 16, 38–9, 139–40
  cryofuels used, 16, 39, 125–36, 143–75, 198
    basic assumptions, 162
    components from space applications, 108, 120
    configuration studies, 162–5
    effect on airframe, 140, 162–3
    R&TD requirements, 121–2, 130–2, 174–5
  operating costs, 168–9
  pollution by, 26–8, 125, 169–74, 196–7
Air transport
  growth trends, 5–6
Alternative fuels
  in air transport, 16, 31, 38–9, 120, 121–2, 123–75
  availability, 12–14
  characteristics compared, *37*, *38*, *139*
  in road/rail transportation, 15–16, 39–40, 124–5, 177–87
  use in transport sector, 14–16, 29–31, 137–40
  factors affecting choice, 15, 35–40
  see also Cryofuels; Liquefied natural gas (LNG); Liquid hydrogen (LH$_2$)
Ariane space project
  liquid hydrogen storage tanks, 96, 106, *107*
  rocket engine, *112*
  testing facilities, 104–5
Automotive transportation
  hydrogen-fuelled, 15–16, 40, 124–5
  see also Road vehicles

Batteries, 30–1
  compared with other energy sources, *37*, *180*
Biomass, 13, 53–5
  conversion to hydrogen, 14, 53, 67
  potential resources, *13*, *52*, 53–5
  world usage, 13, *42*, 53
Biomass-burning electricity generating plants, *52*, 55
  costs, *66*, 67
  in USA, *60*
Biomass-derived fuels, 13–14, 15, 16, 30, 36, 40, 53, 67
  compared with other fuels, *38*
BMW hydrogen-fuelled vehicles, 184

INDEX

Carbon dioxide (CO_2) emissions, 26
  greenhouse effect, 20, 26, *27*, 30, 124, *171*
  historical trends, 19–20, *20*
  by hydrogen production methods, 83
  projected trends, *22*
  reduction target, 20, 28
Carbon/energy taxes, 17
Carbon monoxide emissions, 25
Catalytic converters, 28
Chlorofluorocarbons (CFCs),
  historical trends, *20*
CIS countries
  projected change to cryogenic fuel for air transport, 134, *135*
  see also Russia
Clean energy sources, 41–70
  see also Renewable energy sources
Coal
  price trends, 10
  reserves/resources, *8*, 9, *11*, *12*
  world usage, *8*, *42*
Coal-burning electricity generating plants,
  costs, *66*
Cogeneration projects, 198
Contrails [in tropopause], 26, 27, 173–4, 196
  altitude–temperature effects, *28*, *173*
  cloud coverage affected by, 27–8, *29*
  effect of hydrogen as aircraft fuel, 31, 198
Conventional wisdom energy requirements scenario, 4
Costs
  aircraft operating, 168–9
  electricity generation, 13, *14*, 45, 46, 51, 53, 55, 61–7
  hydrogen-fuelled road vehicles, 185–6
  hydrogen production, *14*, 75, *80*, 82, 84, 185, 192, 193–4
Cryofuels
  aircraft applications, 16, 39, 125–36, 143–75
  auxiliary power unit (APU) supply, 131–2
  effect on airframe dimensions, 140, 162–3
  engine design characteristics, 141, *142*, 143–50
  fuel supply and control system, 131, 145–6, 152–3, 166–8
  refuelling procedure, 119, 132
  compared with other fuels, *37*, *38*, *139*, 140, 162–3, 164
  cooling capacities, 140, 154
  production technology, 85–95
  safety record/requirements, 99–103
  space applications, 104–8, 111–20
  storage and distribution, 96–9
  see also Liquefied natural gas (LNG); Liquid hydrogen (LH_2)
Cryogenic fuel pumps, 131, 145–6, 152
Cryoplane project, 121–2, 126–30, 151
  conclusions/proposals resulting, 133–4
  partner companies involved, *161*
  problems to be solved, 130–2

Daimler-Benz hydrogen-fuelled vehicles, 183–4
DASA
  in Cryoplane project, 121, 126, *161*
  current studies on alternative [aviation] fuels, 161–74
  in international aeroengine study group, 151, *161*, 198
  test facilities, 115–17
Developing countries
  energy demand, 2–4, *42*
Economics
  hydrogen compared with gasoline, 195–6
  hydrogen liquefaction processes, 90–5
  LH_2-fuelled aircraft, 168–9
  renewable energy sources, 61–7
  water electrolysis processes, 78–9
Electricity generation, 7, 9, 13, 30
  costs, 13, *14*, 45, 46, 51, 53, 55, 61–7, 68
Energy carriers, 35–6
  characteristics, 36–7
  selection
    for aircraft, 38–9
    criteria, 37–8
    for road vehicles, 39–40

Energy density
  comparison of various fuels, *37*, 38, 39, *180*
Energy requirements
  Germany,
    transport sector contribution, *23*
  global requirements, 1–4
Energy reserves/resources, 7–9
Environmental aspects, 19–33, 123, 147–8, 169–74, 196–7
Euro-Québec Hydro-hydrogen Pilot Project (EQHHPP), 189–200
  funding arrangements, 199
  future investigations, 199–200
  industrial partners, 191–2
  overview of concept, *190*
  Phase II, 191–5
  Phase III.0, 197–9
  project management, 192
  project phases/milestones, 190–1
Explosions
  factors affecting, 99, 100–1

Flammability range
  hydrogen, 99, 137–8, 171, 196
Fossil fuels, 35
  pollutants from, 20–1, 23–6, *178*
  reserves/resources, 7, *8*
  see also Coal; Gasoline; Kerosene; Natural gas; Oil
Fuel cells, 197, 198

Gasoline
  cost comparison with hydrogen, 195–6
  in dual–fuel vehicles, 181–2
  energy density compared with other fuels, *180*
  NO_x emissions, *178*
Geothermal energy, *13*, 41
Germany
  energy requirements, *23*
  nitric oxide emissions, *25*
  water-electrolysis plants, 78
    potential electricity available, 77
Global energy requirements, 1–4
Greenhouse effect
  and carbon dioxide, 20, 26, 27, *27*, 30, 124

effect of various fuels, 172–3
and methane, 30, 164
and water vapour, 26–7, 31, 171–2, 173
effect compared with that of CO_2, 27, *171*
Ground transportation
  alternative fuels, 177–87
  see also Road vehicles

Household units
  hydrogen generation, 82
Hydride storage tanks, 179
  advantages, 180, 187
  characteristics, *180*, *181*
  costs, 185
  energy density, *180*, *181*, 183, 185
  in road vehicle study, 181–2
Hydrogen
  advantages as fuel, 31, 111, 137–8, 147
  as road transport fuel, 39, 82, 125
  combustion products, *24*, 31, 39, 147
  distribution pipe networks, 75, 96–7
  embrittlement of steel by, 119
  flammability/ignition properties, 99, 137–8, 171, 196
  liquefaction, 86–7, *88*
    effect of process parameters, 92–5
    thermodynamic efficiency, 87–8
  ortho-para conversion, 86, *87*
  production
    carbon dioxide balances, 83
    costs, *14*, 75, *80*, 82, 84
    decentralized installations, 81–3
    energy requirements, 82–3
    energy sources used, 7, 13, 45, 51, *52*, 53
    integration in electricity supply system, 76–9
    methods, 14, 16, 31, 39, 75, 79–81, 85
    size of plant, 89–90
  purification, 86
  see also Liquid . . .; Slush hydrogen
Hydropower, 7
  development/R&TD trends, 55, 68
  electricity production costs, 61, *62*, 68
  future trends, 7, *11*, *12*

Hydropower (cont.)
  hydrogen production using, 7, 13, 189–200
  largest schemes listed, 56
  potential resources, 13, 43, 52, 196
  world usage, 7, 42
Hythane [hydrogen + natural gas], 197–8

Insulation
  cryogenic tanks, 97, 125, 167, 175
International Panel on Climate Change (IPCC), 19

Kerosene
  biomass-derived, 38
  combustion products, 24, 169, 170
  compared with other fuels, 37, 38, 139–42, 139
  price comparison, 168, 169
Kvaerner [hydrogen production] process, 79–81

Lean-burn engines, 151–2, 170–1, 178
Liquefied natural gas (LNG)
  as aviation fuel, 39, 129, 139–40, 143
  combustion characteristics, 147–8
  compared with other fuels, 37, 38, 139–40, 139, 143
  composition, 139
Liquefied oil gas, 142–3
Liquid hydrogen (LH$_2$)
  as aviation fuel, 1, 16, 39, 121–2, 125–6, 129, 135, 139–48, 158–75
  compared with other energy sources, 37, 38, 139, 180
  contamination by solid [nitrogen/oxygen] particles, 119, 132
  method of prevention, 119, 132
  distribution/transportation, 96, 114
  economic feasibility projection, 16
  price, 76, 84, 168, 193
  production
    location of plants, 85
    methods, 16, 86–9
  regulations covering, 103, 133
  in road vehicles, 180–1, 184, 185, 187
  safety considerations, 99–103, 106, 193
  storage, 96, 106, 107
  insulation of tanks, 97, 125
  R&TD studies, 97, 98–9, 198–9
  safety requirements, 101–3
Lockheed hydrogen-fuelled aircraft, 157, 158, 159, 160
Long-term trends, 1–18

MAN buses, 185, 197
Materials research
  cryogenic tanks, 174
Mazda hydrogen-fuelled vehicles, 185
Mercedes-Benz hydrogen-fuelled vehicles, 182
Methane
  as greenhouse gas, 30, 164
  historical trends, 20
  sources, 30, 164
  see also Natural gas
Methylcyclohexane (MCH)
  as hydrogen carrier, 190
  advantages/disadvantages, 194–5
  costs, 193–4

NASA/NACA aeroengine studies, 137, 157–8
Natural gas
  as alternative fuel, 29–30
  combustion products, 24, 30, 164
  composition, 29, 164
  price trends, 11
  production of hydrogen from, 79–81
  costs, 14, 80, 84
  reserves/resources, 7, 8, 11, 12, 134, 143
  world usage, 7, 8, 42
  see also Liquefied natural gas; Methane
Nitrogen oxides (NO$_x$) emissions, 20, 23, 25, 178
  environmental effects, 23, 31, 170
  from hydrogen combustion, 31, 170, 171, 178, 179
  reduction [from engines], 31, 151–2, 170, 178, 198

NK-88 aeroengine
  characteristics, *150*
  combustion chamber, *148*
    characteristics, *149*
  flight trials, 127, 148–9
  LNG operation, 149–50
NK-89 aeroengine, 151
Nuclear power, 9, *11*, *12*, 36, *42*
  see also Uranium

Ocean thermal energy conversion (OTEC), 43–4
  development/R&TD trends, 55, 57
  electricity production costs, 62, *62*
  potential resources, *13*, 44–5, *52*
OECD countries,
  energy demand forecasts, 2–4
Oil
  price limits, 10
  reserves/resources, 7, *8*, *11*, *12*, 138
  world usage, 7, *8*, *42*
Ozone
  trends, 23, *24*

Per capita energy demand, 2–3, *42*
Photochemical smog, 23, 25
Photovoltaic (PV) systems, 53, 58–9
  costs, 65, *65*, 67, 68, 69
Pollutants
  aircraft-derived, 26–8, 125, 169–74
  cryogenic fuels compared with kerosene, 147–8, *170*, *172*, 173
  fossil fuel derived, 20–1, 23–6, 123
  long-lived, 26
  medium-lived, 25
  short-lived, 23–5
Population growth, 2
Price forecasts
  conventional fuels, 10–12
Public transport
  cost breakdown, 197
  hydrogen-fuelled bus trials, 185, 197–8

Real cost pricing [of energy], 70
Regulations
  liquid-hydrogen systems, 103, 133

Renewable energy sources, 9, 12–14, 36
  costs, 13
  development/R&TD trends, 55–61
  economics/efficiency trends, 61–7
  potential resources, *11*, *12*, *13*, 43–55, 67–8
  see also Biomass; Hydropower; Ocean . . .; Solar . . .; Tidal . . .; Wave . . .; Wind energy
Road vehicles
  alternative fuels, 15–16, 39–40, 177–87
  hydrogen-fuelled, 15–16, 40, 124–5, 178–87
    availability, 183
    BMW designs, 184
    components from space applications, 120, 187
    costs, 185–6
    Daimler-Benz designs, 183–4
    dual-fuel design, 181–2
    emissions comparison, *179*
    EQHHPP studies, 197–8
    Mazda designs, 185
    Mercedes-Benz designs, *182*
    obstacles to be overcome, 186
    on-board hydrogen storage, 179–81
    pilot scheme, 181–2
    R&TD required, 121, 183
Rocket engines, *108*, *112*
  test facilities, 104–5, *113*, 115–17
    accidents occuring, 118–19
Russia
  natural gas resources, 134, 143
  oil resources, 138
  projected change to cryogenic fuel for air transport, 134–6

Safety aspects
  hydrogen-fuelled aircraft, 130, 142
  hydrogen gas, 99, 126
  liquid hydrogen storage/distribution/handling, 99–103, 106, 193
  rocket engine test facilities, 115–16
Samara State Scientific and Production Enterprise (SSSPE)
  aeroengine studies, 138

Samara State Scientific and Production Enterprise (SSSPE) (*cont.*)
  see also TRUD aeroengine studies
Slush hydrogen, 89
Solar energy, 13, 51–31
  electricity production costs, *62*, 64–7
  potential resources, *13*, 51–2, *52*
Solar thermal electricity production, 53, 59, 64–5
  generation costs, 9, *64*
Space applications, 104–8, 111–20
  components suitable for air/ground transportation, 108, 119–20, 187
Steel production
  hydrogen proposed as reducing agent, 198
Supersonic aircraft, 158, *159*

Targeted efficiency (TE) energy requirements scenario, 3, 4, 12
Targeted growth (TG) energy requirements scenario, 3, 11
Temperature trends
  historical data, *21*
  predictions, *21*
Tidal energy, 46–7, 57, 68
  electricity production costs, *62*, 63
  potential resources, *52*
Tonne of oil equivalent (toe)
  energy content, 3
Transport sector
  alternative fuels, 14–16
  fuel requirement trends, 4–7
  see also Air transport; Road vehicles
TRUD
  aeroengine studies, 138, 140–50
  in international study group, 151, *161*
TU–155 experimental aircraft, 126–30
  components, 128–9
  conclusions/proposals resulting, 133–4
  general description, 126–8
  LNG-fuelled operation, 129, 149–50

refuelling procedure, 119, 132
test facilities, 129–30
TU–156 LNG-fuelled aircraft, 134–6, 151
Tupolev Company,
  LH$_2$-fuelled experimental aircraft, 126–30

UN Framework Convention on Climate, 20, 21, 28
Uranium
  price trends, 11

VFW-designed aircraft, 159, *160*
Vulcain rocket engine, *112*
  testing facilities, 104–5

Water electrolysis, 14, 31, 39
  costs, *14*, 75, *80*, 84
  economics of hydrogen production using, 78–9, 192, 193
  integration in electricity supply system, 76–8
  solid polymer electrolyte (SPE) hydrogen generator, 82
Water vapour
  as combustion product, 26, 31
  emissions from aircraft, 26, 27, 173–4, 196
  evaporation rates, 196
  and greenhouse effect, 26–7, 31, 171–2, 173, 196
Wave energy, 45–6, 57
  electricity production costs, 62–3, *62*
Wind energy, 13, 47–51
  development/R&TD trends, 57–8, 68
  electricity production costs, *62*, 63, 68, 69
  potential resources, *13*, 47–8, *50*, *52*

Zero-emission road vehicles, *179*
  legislation requiring, 39, 186

Index compiled by Paul Nash